Mesoscopic Nuclear Physics

From Nucleus to Quantum Chaos to Quantum Signal Transmission

Mesoscopic Nuclear Physics

From Nucleus to Quantum Chaos to Quantum Signal Transmission

Vladimir Zelevinsky
Michigan State University, USA

Alexander Volya
Florida State University, USA

NEW JERSEY • LONDON • SINGAPORE • BEIJING • SHANGHAI • HONG KONG • TAIPEI • CHENNAI • TOKYO

Published by

World Scientific Publishing Co. Pte. Ltd.
5 Toh Tuck Link, Singapore 596224
USA office: 27 Warren Street, Suite 401-402, Hackensack, NJ 07601
UK office: 57 Shelton Street, Covent Garden, London WC2H 9HE

Library of Congress Control Number: 2022945022

British Library Cataloguing-in-Publication Data
A catalogue record for this book is available from the British Library.

MESOSCOPIC NUCLEAR PHYSICS
From Nucleus to Quantum Chaos to Quantum Signal Transmission

ISBN 978-981-126-314-9 (hardcover)
ISBN 978-981-126-315-6 (ebook for institutions)
ISBN 978-981-126-316-3 (ebook for individuals)

For any available supplementary material, please visit
https://www.worldscientific.com/worldscibooks/10.1142/13049#t=suppl

Desk Editor: Nur Syarfeena Binte Mohd Fauzi

Typeset by Stallion Press
Email: enquiries@stallionpress.com

Contents

Chapter 1

Introduction

In the beginning of the 20th century, nuclear physics started with the discovery of a positively charged small object in the center of every atom, the *nucleus*. Gradually, it was found that the nucleus consists of *nucleons* — positively charged protons and neutrons with no electric charge. These particles interact through strong *nuclear forces*, very similar for protons and neutrons, keeping the nuclear constituents together inside a volume by five orders of magnitude smaller than the size of the atom defined by the electronic cloud. Many nuclei of the same chemical element live in various modifications (*isotopes*) which differ by the number N of neutrons for the same electric charge defined by the number Z of protons. The chemistry is mainly determined by the electrons and the atomic level architecture of molecules and crystals. At typical nuclear density, the intrinsic quark–gluon structure of nucleons is not revealed explicitly so that its study requires huge accelerators or extremely refined experimental techniques.

The 20th century demonstrated broad applications of nuclear physics in peaceful and military technology. The harvesting of specific isotopes is one of the main goals of the new nuclear accelerators. The abundance of natural isotopes in the world is a crucial ingredient for modern theories of the evolution of the universe. The nuclei also conceal keys to the understanding of the charge asymmetry of the universe, many properties of elementary particles, their interactions and decays.

We still do not know how far the chart of nuclear isotopes can stretch for elements heavier than oxygen or in the direction of superheavy elements. More than 7000 isotopes may exist according to our extrapolations, while hardly half of them were observed in the laboratory. Figure 1.1 gives an idea of the known isotopes and predictions for the boundaries of their existence,

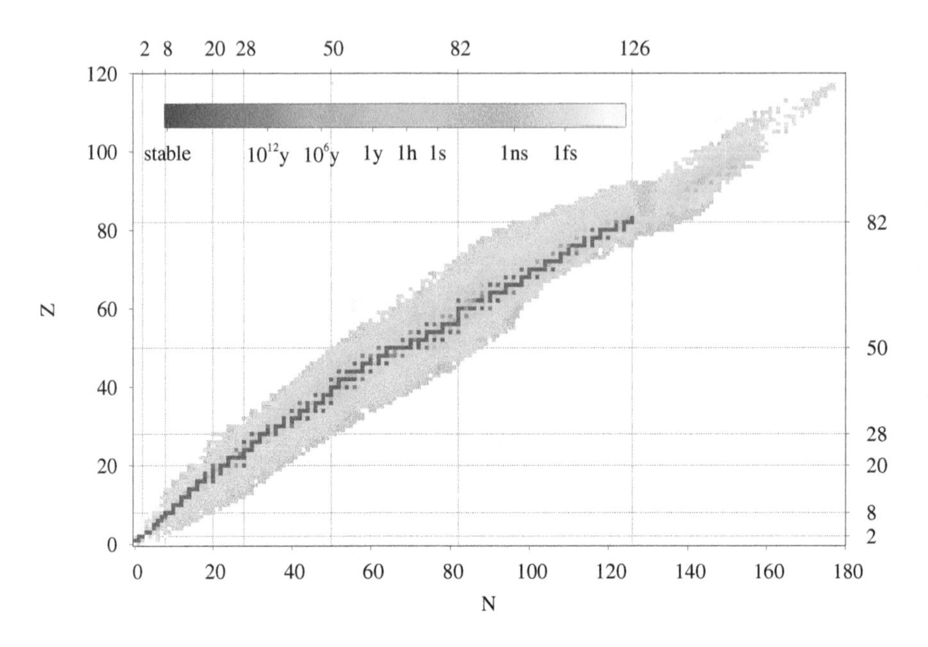

Fig. 1.1. Chart of nuclides color coded by lifetimes.

the so-called *drip lines* where the extra nucleons cannot be kept bound by the nucleus.

From a structural viewpoint, the nucleus is a self-bound system of strongly interacting particles. It shares common features with other natural systems of such type: complex atoms, complex molecules, including biological, atomic clusters. Moreover, there are many man-made objects of this class, for example, artificial condensed matter systems on a micro- and nano-scale, atoms in traps, etc. Future quantum computers will contain working elements (*qubits*) interacting in the computational processes. There are general features characterizing the systems of this kind which can be thought of as members of a special class — *mesoscopic*, intermediate between the microscopic and macroscopic worlds. With clear understanding that the boundaries between these communities are quite conditional and blurred, it is still possible to mention the main regularities common for mesoscopic objects.

In this intermediate world, the inhabitants consist of interacting constituents usually ruled by the laws of quantum mechanics. The intrinsic interaction determines observable characteristics of the system that can be fully autonomous like an isolated atom or nucleus, or subject to external fields as a quantum dot in modern solid state physics, cold atoms in a trap, or man-made constructions for quantum signal transmission. The number of constituents is typically large enough for the manifestation of statistical laws. At the same time, mesoscopic systems are small enough to provide an opportunity to study theoretically and experimentally individual quantum states. This situation creates the wealth of scientific ideas and practical applications.

The interacting constituents being dressed by the interaction effects in a many-body environment become *quasiparticles* along with quasiparticles of another type — quanta carrying the interaction signals. The first fundamental effect of the interaction is the formation of the self-consistent *mean field* whose symmetry defines the quantum numbers of quasiparticles propagating in this field. The residual interactions between the quasiparticles are responsible for *collective* effects, typically waves propagating through the system. In many situations, the collective effects are sufficiently strong to induce the reshaping of the mean field leading to phase transitions, superfluidity and superconductivity. The collective modes are usually not stationary states of the system, they have a finite lifetime and undergo *damping* — decay into complicated (incoherent) many-body excitations.

With increasing excitation energy, the density of quasiparticle states sharply grows strongly amplifying the role of the residual interactions. The stationary states become exceedingly complicated combinations of many excited quasiparticles. This is very similar to the heating of the system with growth of temperature. It is necessary to stress that thermalization occurs even in a self-supported system with no heat bath, being driven solely by intrinsic interactions. As a result, the excited system may reach some kind of equilibrium that can be compared to the predictions of standard statistical ensembles. In nuclei this stage can be juxtaposed to the Niels Bohr picture of a *compound nucleus*, see the renown Fig. 1.2. The later stage of the development could be decay into continuum if the excitation energy exceeds the decay threshold; when energetically allowed, decay can happen before full equilibration.

The equilibrium stage at sufficient excitation energy allows for another description — from the viewpoint of *quantum chaos*. As a result of interactive mixing, the original few-quasiparticle states become extremely mixed and lose their individuality. Very complicated stationary states at close energy are deprived of their own face, they all have very similar observable properties as was formulated long ago by Percival [180]. The high density of states with the same global quantum numbers can be a source of

Fig. 1.2. Bohr's compound nucleus idea [41]. Illustration by N. Andreeva.

the enhancement of certain physical phenomena, such as violation of global symmetries or quantum tunneling. The mathematical limit of this situation is given by the *random matrix theory* where only the most general properties of quantum dynamics remain in force dividing mesoscopic systems into symmetry classes.

Another angle of looking at an excited by an external signal mesoscopic system is related to its time development from a relatively simple state formed immediately after the initial excitation, for example absorption of a neutron or gamma quantum by a nucleus. The further process leading to the formation of a compound system is in fact the road to thermalization and equilibration without an external thermostat. It can be either cut short along the road (preequilibrium process) or ended by the evaporation from the reached thermal equilibrium.

Our text is not a textbook in nuclear physics. It is devoted to the atomic nucleus as a bright representation of mesoscopic quantum physics. We start with the concise reminder of main properties of complex nuclei. Here we introduce the standard terminology of nuclear science; the necessary details will be added in further chapters as needed. Much more information on nuclear physics can be found in various books including the textbook of the authors on *Physics of Atomic Nuclei* [265]. Main mesoscopic features of nuclei will be reviewed in Chapter 2, while Chapter 3 will show the basic statistical characteristics of the nucleus as a many-body quantum system. Chapter 4 will demonstrate the interrelation of those features and the ideas of quantum chaos. Chapter 5 is based on the results of the most detailed description of the nuclear quantum states and their mesoscopic properties, including the dynamics of the time development of an excited nucleus, with the help of the shell-model diagonalization (exact solution of the many-body quantum problem in the limited Hilbert space). Important aspects of nuclear mesoscopic physics related to the openness of the excited nucleus and coupling through continuum to the outside world are presented in Chapter 6. Finally, Chapter 7 shows few examples of the same approach applied to non-nuclear mesoscopic systems with the analogy of a nuclear reaction to a general process of the quantum signal transmission through a complex quantum arrangement, the necessary step on the road to quantum computing.

We are extremely grateful for friendly collaboration and discussions to N. Auerbach, G.P. Berman, F. Borgonovi, P.F. Bortignon, P. von Brentano, R.A. Broglia, B.A. Brown, Bui Minh Loc, P. Cejnar, G.L. Celardo, V.V. Flambaum, Ya.S. Greenberg, M. Horoi, F.M. Izrailev, L. Kaplan, S. Karampagia, D. Kusnezov, B. Lauritzen, C.H. Lewenkopf, S. Mizutori, S. Reimann, L.F. Santos, R.A. Sen'kov, V.V. Sokolov, Ch. Stoyanov, and S. Åberg. This work gave rise to the dissertations and further promising research of V. Abramkina, N. Ahsan, J. Armstrong, N. Frazier, M. Ghita, K. Kravvaris, D. Mulhall, G. Shchedrin, S. Sorathia, A. Tayebi, D. Volya and A. Ziletti. The participation of undergraduate and high-school students A. Berlaga, T.N. Hoatson, C. Merrigan, A. Renzaglia, J. Spitler, and J. Wang was an important step in their further scientific development.

Chapter 2

Nucleus as a Mesoscopic Object

2.1. Strong, Electromagnetic, and Weak Interactions

Nuclear forces responsible for the bound states of nuclei and numerous nuclear reactions are indeed *strong* as compared, for example, to Coulomb, and more general, electromagnetic forces playing the main role in condensed matter and therefore in physics of practically all phenomena in everyday life. In the first rough approximation, neglecting electromagnetic effects, we can combine protons and neutrons in one family of *nucleons*. This leads to introduction of the new quantum number of *isospin* T as a label in the family of strongly interacting particles and its permutational symmetry. In this approximation, protons and neutrons belong to the simplest representation of the SU(2) group with isospin $T = 1/2$; in nuclear physics by tradition we ascribe the isospin projection $T_3 = +1/2$ to the neutron and projection $T_3 = -1/2$ to the proton (in particle physics the tradition is opposite). The main nuclear forces are *isospin-invariant* but the presence of the Coulomb interaction makes the binding energies and the reaction behavior different for protons and neutrons.

Strong forces in the nucleus have a short distance of action determined by the Compton wave length $\hbar/(m_\pi c) = 1.4$ fm of the lightest particles, the pions, that can serve as carriers of the nuclear force. Heavier mesons and other short-lived particles are responsible for the forces (including repulsion) at even smaller distances. The strong forces lead to the binding energy per particle almost constant along the nuclear chart, 6–8 MeV, with exception of the lightest nuclei where all nucleons are essentially at the surface, and the heaviest nuclei where the electrostatic repulsion between the protons becomes too powerful. The life of the superheavy nuclei ends either by fission into much lighter stable nuclear fragments or by a sequence of decays with emission of very strongly bound helium nuclei (alpha-particles) or even heavier clusters.

The strong attraction and electrostatic repulsion together determine the charge and matter distribution inside the nucleus. Due to the competition between this repulsion and surface tension, the superheavy nuclei are expected to become absolutely unstable at the charge close to $Z \approx 140$. The interaction of an excited nucleus with the quantized electromagnetic field may generate the *radiation* of photons with the transition of the excited nucleus to one of the lower energy levels. As the photon wavelength in such a transition is greater than the nuclear size, $kR \ll 1$, the radiation of low multipolarities, mainly dipole (electric, E1, and magnetic, M1) and electric quadrupole, E2, is usually taking place. Strictly speaking, just the ground state of the nucleus can be really stable; however, the radiative transitions are significantly slower than the typical time of a fast nuclear reaction. The probabilities of radiative transitions are regulated by the standard *selection rules* — conservation of energy, angular momentum (total nuclear spin) and, approximately but with high precision, parity and isospin. In a reaction between colliding heavy nuclei, the total angular momentum of a temporary formed double system can reach values of the order of 70 or 80. Then the system and its individual partners can radiate a long ladder of consecutive electromagnetic transitions, typically of electric dipole character. Such transitions bring the nucleus to low-lying states and eventually to the ground state.

The nuclei occupy an important corner of the current *standard model* of elementary particles and their interactions, strong (proper nuclear), electromagnetic, and weak. Many fundamental discoveries leading to the standard model were made on the nuclear material. The nuclei with not the most energetically favorable ratios between proton and neutron numbers improve their stability through beta-decay or electron capture generated by the *weak interaction* with emission of *neutrinos*. The characteristic times for weak decays cover a big interval, from milliseconds to the time comparable to that of the Earth existence. Such decays in astrophysical conditions are mainly responsible for the formation of the valley of stability on the nuclear chart and for the observed abundances of chemical elements. In distinction to strong and electromagnetic forces, the weak interactions violate spatial inversion symmetry $\mathcal{P}$ (parity conservation) and charge conjugation symmetry $\mathcal{C}$ while their product $\mathcal{CP}$ is still approximately conserved being broken only in rare meson processes with the violation of time-reversal symmetry $\mathcal{T}$ in such a way that the full $\mathcal{CPT}$ symmetry stays invariant as the fundamental property of the relativistic quantum field theory.

Neutrino physics acquired a special value after discovery of oscillations between different types of neutrino. This discovery reveals the non-zero neutrino mass (less than 1 eV but still unknown). In relation to many astrophysical processes, including the stellar evolution, the role of neutrino may be decisive. The ongoing search for exotic nuclear processes, such as the neutrinoless double beta-decay, will decide the problem of nature of the neutrino as an elementary fermionic particle (Dirac or Majorana) [23].

The subject of our interest, mesoscopic features of a nucleus, is closely linked to all types of intrinsic interactions. Strong forces keep nucleons together creating the mean nuclear field with certain shape, the corresponding spatial symmetry, renormalized nucleon properties, and a characteristic binding energy. The interaction remnants not reducible to the static mean field lead to a possibility of collective motion, shape fluctuations and evolution, with the growth of excitation energy, to a very dense and irregular web of nuclear energy levels below decay thresholds. Above thresholds, the former stationary levels become *resonances* seen in various reactions with the widths gradually increasing along with the distance from thresholds. With the growing level density, the interactions lead to the emergence of statistical features, thermalization and quantum chaos, one of our main subjects in this text. The electromagnetic and weak transitions provide the signals of this evolution that proceeds, with the increase of the level density from the vicinity of the ground state up, in the direction of chaotization of nuclear dynamics.

2.2. Single-Particle Degrees of Freedom

We need to introduce an appropriate language for the description of nuclear quantum states. Assuming the existence of a static mean field, in the zero approximation, we put independent protons and neutrons into single-particle orbitals according to the Pauli exclusion principle. Here we do not discuss the formal path to the derivation of the appropriate mean field from original nucleonic interactions but just claim that this field does exist and can be theoretically derived (or fitted phenomenologically) in such a way that it describes reasonably well, but still approximately, the general scheme of appropriate single-particle states and the order of their filling with increase of energy. A good phenomenological example, quite successful, especially for relatively light nuclei, is given by the Argonne potentials [254]; later versions [184] included also three-body forces. A recent more sophisticated approach is based on the energy density functional [198]. But one can even start with the oversimplified model of the trivial three-dimensional harmonic oscillator field, spherical or deformed.

The properties of the nucleus as a whole in the mean-field approximation appear as a result of summation of individual contributions of independent particles. Here we would better call them *quasiparticles* as their single-particle characteristics are modified in a self-consistent way by interactions in the presence of other particles. These properties include effective masses, gyromagnetic ratios, etc.; in some cases one needs to introduce effective charges, different for protons and neutrons. The single-particle wave functions are the eigenstates of the effective energy functional sequentially populated in agreement with the Fermi statistics. The corresponding philosophy is borrowed from the Landau's [173] Fermi-liquid theory and the Kohn–Sham energy density functional [136] being adjusted for a mesoscopic system as the nucleus by Migdal and his school [157].

Equipotential surfaces of the mean field define the mean nuclear shape. In spherical nuclei, the single-particle states are characterized by the integer orbital momentum ℓ, corresponding parity $(-)^{\ell}$, and the total single-particle angular momentum $j = \ell \pm 1/2$. The spin–orbit coupling is strong in nuclei being approximately described in the mean-field Hamiltonian as a sum of single-particle contributions,

$$H_{\ell s} = -\sum_{a} h(r_a)(\vec{\ell} \cdot \mathbf{s})_a, \tag{2.1}$$

driving down the level $j = \ell + 1/2$ of the spin–orbit doublet and pushing up the level $j = \ell - 1/2$. The positive form-factor function $h(r)$ is mainly concentrated close to the nuclear surface as in the interior there are no directions singled out for the spin orientation. In distinction to atomic systems, where the spin–orbit effects of relativistic origin are weak (and of the opposite sign compared to eq. (2.1)), the nuclear spin–orbit coupling comes from strong forces of meson exchange. As it is natural to expect, the mean density in the nuclear interior is approximately constant, the nuclear volume is proportional to the number A of nucleons with the radius growing in average as $A^{1/3}$. The typical spin–orbit strength $h \approx 20/A^{2/3}$ MeV clearly shows the surface behavior, an example of the obvious mesoscopic effect.

The radial shape of the mean-field potential well is the secondary effect from the mesoscopic point of view. Traditionally, a harmonic oscillator field is frequently used for qualitative estimates,

$$U(\mathbf{r}) = \frac{1}{2} m\omega^2 \mathbf{r}^2. \tag{2.2}$$

This cannot be exact because of the unrealistic asymptotic radial behavior of single-particle orbitals but it is convenient as a basis for the diagonalization of the total Hamiltonian of interacting bound nucleons. For the description of processes coupled to the continuum, specific corrections could be necessary. The primitive oscillator scheme with spin–orbit coupling and expected in this scheme *magic numbers* of completely filled shells are shown in Fig. 2.1. This figure also includes the interrelation to other simple mean-field versions. A spherical square well is perhaps the simplest potential with realistic remote asymptotics of nucleon wave functions describing the physics in the continuum; the popular *Woods–Saxon potential* predicts more realistic sequences of single-particle shells and magic numbers for spherical nuclei with fully occupied shells [199, 256]. The whole picture

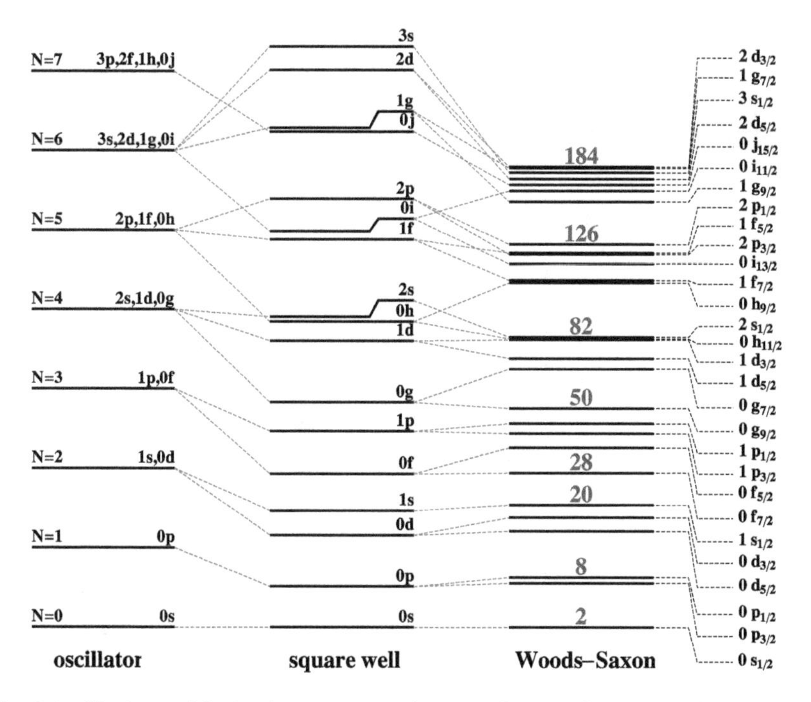

Fig. 2.1. Single-particle level structure and nomenclature of corresponding quantum numbers in the traditional mean-field models; for the Woods–Saxon potential with spin–orbit coupling, magic numbers of filled shells are indicated.

may become less realistic for short-lived nuclei closer to the drip line where the influence of the continuum becomes an important factor, especially for loosely bound orbitals.

Strictly speaking, the shape of the mean field is defined only in the *intrinsic* frame with the axes attached to the body of the nucleus and with frozen rotational degrees of freedom. The simplest description of the mean field with quadrupole deformation can be made in terms of an *anisotropic* harmonic oscillator. In the potential (2.2) written now in terms of the body-fixed variables, we can simply use different frequencies along different intrinsic axes. The relation between the lab frame and the body frame is given through the matrix elements of finite rotations parameterized by rotation angles. This is strictly valid only in the case of a rigid body [265]. For soft nuclei with the easily excitable surface and less certain definition of the intrinsic frame, it is better to use the lab frame with the well-defined total nuclear spin; the deformation effects will show up in consideration of nuclear excitations and form-factors. The deformation properties will reappear later in the discussion of the density of states and some special nuclear observables.

Adding the spin–orbit coupling and the potential proportional to $\vec{\ell}^2$ that makes, in agreement with data, the bottom of the potential flat, we come, Fig. 2.2, to the *Nilsson scheme* for the axially symmetric quadrupole deformation [172]. As a function of the deformation parameter, the single-particle j-levels are split according to the square of the projection $j_z \equiv m$ onto the internal symmetry axis; the time-reversal symmetry keeps the orbitals $\pm m$ degenerate. The right-hand side of the Nilsson scheme shows that, in the beginning of an oscillator shell, the prolate axial deformation is energetically favorable, while in the end of the shell the oblate deformation becomes preferable. This is strictly correct within one split j-orbital, in agreement with the actual clear preponderance of the prolate shape in stable deformed nuclei. One should remember that the single-particle quantum number ℓ is not exact at the deviation from sphericity; in a sense, mixture of levels with $\Delta\ell = \pm 2$ is a signature of quadrupole deformation [126] of the mean field. Levels from upper shells with higher ℓ and j being split by the square of the axial projection come sharply down with increase of deformation and approach the ground state frequently again making the prolate shape energetically preferable. As a result, the majority of stable deformed nuclei have a prolate shape; among unstable nuclei, the fraction of

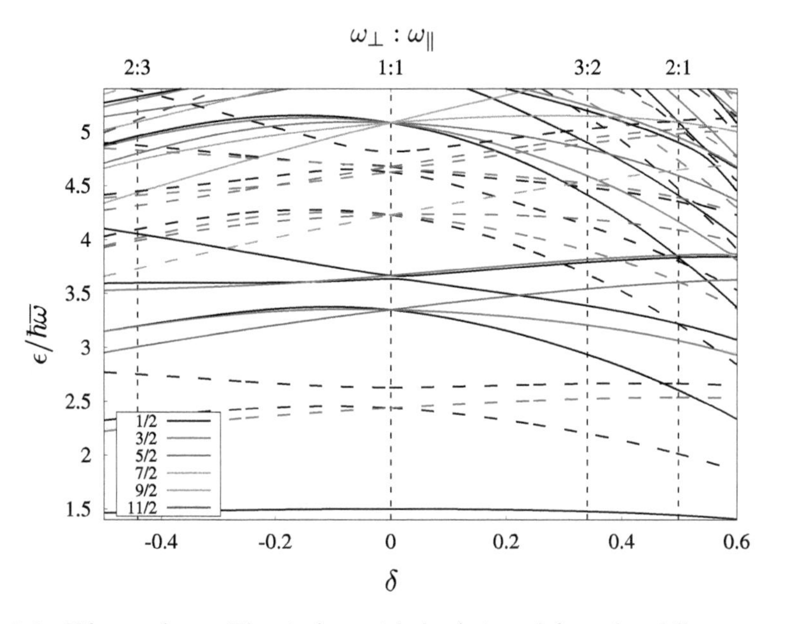

Fig. 2.2. Nilsson scheme. The single-particle levels in a deformed, axially symmetric harmonic oscillator potential, shown as a function of deformation; see Ref. [265] for details.

oblate deformation is higher. There are also regions of the nuclear chart (for example, osmium isotopes) without axial symmetry where the transition from deformed to spherical near-magic nuclei proceeds through tri-axial shapes.

We can consider the mean nuclear field taken in the vicinity of the ground state either as a result of self-consistent calculations based on a certain energy density functional or with an *ab initio* formulation using meson-mediated vacuum interactions. In practice, some phenomenological corrections might be useful in order to obtain a better description of the data. As an average picture of nucleon life, the mean field concentrates regular quantum-mechanical features of nuclear dynamics, in contrast to collective and random (stochastic) features brought in by quasiparticle interactions. This provides a possibility to measure the collectivity and chaoticity by comparison to the mean field picture. The mean field may evolve with excitation energy in a self-consistent way, along with general stochastization of the dynamics. This evolution is not entirely chaotic as it includes collective excitations of the system.

2.3. From Mean Field to Fermi-Liquid

Now we move to a more specific discussion of nuclear many-body characteristics. A typical nuclear system is a self-bound conglomerate of interacting nucleons — neutrons and protons (we do not go deeper to quarks and gluons that becomes necessary for the nuclear collisions at very high energy [204]). As mentioned earlier, the main forces between the nucleons are due to the exchange of mesons with the range defined by the lightest mesons, pions, $r_0 \sim \hbar/m_\pi c \approx 1.4$ fm, that is small compared to the typical nuclear radius approximately growing with the mass number A as $R \sim r_0 A^{1/3}$. This shows that the nuclear density in the interior is approximately constant, a sign of *saturation* of nuclear forces. The presence of the surface leads to the surface tension, a consequence of non-saturated forces between nucleons in surface orbitals. The interplay of surface tension and Coulomb repulsion between the protons is responsible for the fission instability of the heaviest nuclei. At short distances, the nucleons interact through exchange of heavier mesons. This part of the interaction contains repulsive forces, mainly due to the quark core.

The nucleonic interactions are strongly influenced by the Fermi-statistics. Due to the relatively large mean interparticle distances, typically it is possible to consider nucleons as pure fermions although, strictly speaking, having a complicated substructure of quarks and gluons, a nucleon is not a standard fermion. At energies we are going to consider, and for typical nuclear densities, this feature of compositeness can practically be neglected. Therefore, we usually treat the nucleus as a self-supported drop of the two-component Fermi-liquid. At not very high excitation energy, the equilibrium density is constant in the interior as an indication of nuclear *incompressibility*.

The Coulomb repulsion makes energetically favorable having a certain relation between numbers of protons and neutrons. Neglecting electrostatic forces and the small neutron–proton mass difference, we would have $N = Z$ in equilibrium. This is a consequence of the isospin invariance of nuclear forces when one can consider the neutron and the proton as two substates of the nucleon with isospin 1/2. The electromagnetic interactions (as well as quite small parts of nuclear forces) violate isospin invariance. As a result, the line of nuclear stability prefers isotopes with $N > Z$ while the nuclear species deviating from this line undergo weak decays returning to this line in agreement with the picture of nucleosynthesis of stable elements in stellar environment [195]. Currently, the properties of many unstable nuclei with relatively long lifetime determined by the weak interactions are well studied.

The main purpose of several modern nuclear laboratories is to delineate the filling of the nuclear chart by advancing from the line of stability in the direction of the drip lines.

The creation of the *mean field* is the simplest and most general effect of interparticle interactions. The nucleonic quasiparticles ("dressed" nucleons) are distributed over certain orbitals defined by the symmetry of the mean field, spherical or deformed. As mentioned earlier, some properties of quasiparticles, such as effective masses, magnetic moments, etc., can be different from those for original nucleons being dependent on the occupied orbital in a nuclear environment. Adding the residual interactions between the quasiparticles we come to the Landau–Migdal picture of a nucleus as a drop of Fermi-liquid [157].

The quantum numbers of single-particle orbitals are defined by the shape of the mean field. As residual interactions influence the static mean-field picture, the equilibrium nuclear shape can change from nucleus to nucleus that was illustrated by the Nilsson diagram of Fig. 2.2. The main types of the residual interaction are pairing and multipole–multipole, first of all quadrupole-quadrupole, forces. As a result of their interplay, the nuclei close to filled shells (*magic nuclei*) are spherical while at some degree of filling the upper shells, the deformed shape becomes energetically favorable. It is useful to recall that the discussion of the nuclear shape refers to the *intrinsic frame* with axes defined by the internal nuclear geometry. This frame is independent of the *laboratory frame* where the nuclear orientation in space is given by the quantum-mechanical eigenfunctions $|JM\rangle$ of the conserved total angular momentum. In the limit of the well-defined rigid body, the angular wave function of the nucleus as a whole is given by the standard function D_{JM} of Euler angles [265] which describe the nuclear orientation in the lab space.

Given the shape of the mean field, a nucleonic configuration as a whole is defined by the *partition* of the nucleons over single-particle orbitals occupied in this mean field according to the Fermi statistics (Pauli exclusion principle). An important part of the classification of orbitals is the presence of spin–orbit coupling mentioned above, Eq. (2.1). In contrast to the small atomic spin–orbit splitting of electron levels, this coupling is strong defining the closed shells and corresponding magic numbers.

Already the simplest harmonic oscillator plus the spin–orbital coupling scheme, Eq. (2.2), of nuclear orbitals with the large degree of degeneracy leads to a reasonable picture of nuclear shells that reproduces the known magic nuclei (the scheme can be modified for nuclei outside the valley of stability). We notice that the single-particle levels come naturally in groups of the same parity while the levels of opposite parity are shifted down from the higher shell through spin–orbital splitting and destroy the previous pure parity. The markedly appearing shell structure allows us to assume that the mean field accumulates the most regular elements of nuclear dynamics, even if the details of the mean field can be different in different approaches. The mean field in practice can be phenomenological on derivable from the energy density functional, for example of the Skyrme type [219]. The concrete choice can influence details of spectroscopy and low-energy reactions but does not radically change the main qualitative trends. The main difference of the Fermi-liquid picture from the mean field is the presence of residual interactions.

2.4. Residual Interactions

The mean-field partitions are not stationary quantum states. They just provide a convenient complete basis for the description of actual nuclear many-body eigenstates. The way from the mean field (free quasiparticle gas in this field) to realistic nuclear states and observables is governed by residual interactions. It is convenient to express the total nuclear Hamiltonian in the mean-field basis with single-particle orbitals $|\nu)$ where the shortened label ν (that frequently will be substituted by numbers $1, 2, \ldots$) combines all single-particle quantum characteristics. In practical calculations, it may be convenient to use the basis of a spherical field, where $|\nu)$ contains the isospin projection (proton or neutron), main quantum shell number n, orbital momentum ℓ that defines parity $(-)^{\ell}$ of a spherical single-particle level, total angular momentum $j = \ell \pm 1/2$ with the larger of the two values of j having lower single-particle energy ϵ_{ν}, and the projection $j_z = m$ on the laboratory quantization axis. In the mean-field basis, this part of the total Hamiltonian can be taken diagonal in terms of the corresponding fermionic creation and annihilation operators,

$$H_0 = \sum_{1} \epsilon_1 a_1^{\dagger} a_1. \tag{2.3}$$

The *two-body* residual interaction can be written in the operator form as

$$H_{\text{int}} = \frac{1}{2} \sum_{1234} V(12;43) a_1^{\dagger} a_2^{\dagger} a_3 a_4. \tag{2.4}$$

One can also use the antisymmetrized form with respect to the subscripts 1 and 2 and change the prefactor 1/2 to 1/4. The matrix elements V satisfy the conservation laws of angular variables J and M of the initial and final pairs. Those conservation laws, as well as the approximate isospin conservation, are expressed in terms of corresponding Clebsch–Gordan coefficients in angular and isospin spaces [265]. By explicit commutation and recoupling of angular and isospin quantum numbers (the so-called *Pandya transformation* [178]), one can transform the Hamiltonian to the particle–hole channel in the form $(a^\dagger a)(a^\dagger a)$.

In all practical calculations, the orbital space of the interaction Hamiltonian is truncated (finite sums over the orbitals $1,2,3,4$ in Eq. (2.4)). Therefore, what enters here is usually not the original interaction of free nucleons but the effective interaction for the given mean field including the renormalization of vacuum forces by the presence of the nuclear medium, as well as virtual admixtures of the shells lost in the process of truncation. In practice, this effective interaction contains parts fitted by the comparison to the limited practically available experimental data. This renormalization does not change the exact quantum numbers of the states.

Currently, there exist modern approaches to the derivation of the effective nuclear Hamiltonian based on the chiral effective field theory [84, 153]. We do not go into details here as anyway all reasonable nuclear effective Hamiltonians should give the results comparable to the semi-phenomenological shell model that successfully describes low-lying empirical information and therefore can be reliably used for the analysis of statistical nuclear properties. We just mention that, strictly speaking, the chiral field theory with its phenomenological constants reflecting the influence of the short-range part of the dynamics, leads to *many-body* interactions rather than to two-body only as in Eq. (2.4). The many-body effects can be mimicked by the dependence of two-body matrix elements on local nuclear density [83]. This should not change (but can accelerate) the general trends leading from regular dynamics to chaos that are of our main interest.

2.5. Nuclear Pairing

Among residual nuclear interactions, a special role belongs to *pairing*. It was known long ago, starting with the introduction of the basic shell model [105], that the nucleons in the upper open shells reveal the preference with respect to occupying pairwise the degenerate time-reversed single-particle orbits, for example with angular momentum projections $\pm m$. The pairing energy provides the characteristic mass differences between even–even, odd–even and odd–odd nuclei and guarantees the ground-state spin-parity quantum numbers $J^{\Pi} = 0^{+}$ in the even–even case. The existence of the attractive pairing energy is essentially the reason why ^{235}U can fission being excited by a slow neutron while ^{238}U cannot [119].

Assuming the pairing, one can, in many cases, predict the quantum numbers (quite reliably) and magnetic moments (approximately) of the ground states in the odd-A neighbors as coming from the last unpaired nucleon. The underlying algebraic structure of paired states was established by Racah [188, 265]. After the success of the based on pairing BCS (Bardeen–Cooper–Schriffer) theory [25] of superconductivity in metals, A. Bohr, Mottelson and Pines suggested [42] that nuclear pairing has a similar nature carrying a mesoscopic embryo of superconducting correlations. The full microscopic theory of pairing effects in nuclei was constructed by Belyaev [30], while the later development along with the discussion of further perspectives is broadly covered in the special volume [52] devoted to the 50th anniversary of the idea of pairing in nuclei.

Being in fact a part of the residual interaction, nuclear pairing can be formally considered as a certain deformation of the mean field (in *gauge space* with the pair condensate characterized by a certain phase). This is a regular effect that influences all nuclear observables. Apart from the even–odd mass difference (a typical pairing energy, Fig. 2.3), pairing changes the characteristics of nuclear excitations, moments of inertia, probabilities of various processes, and creates a gap 2Δ in the single-particle spectra of even–even nuclei. The formally evaluated spatial correlation length of a BCS pair is greater than the nuclear size.

Due to the discrete nature of the quasiparticle spectra, there is no Cooper instability of the normal Fermi surface; the pairing interaction has to be stronger than a certain limit to lead to the superconducting gap. Therefore, one cannot fully transfer the theory of macroscopic superconductivity to finite nuclei, in spite of many similar features. With increase of the excitation energy, similarly to the heating of a superconductor, the broken by thermalization pairs block the superconducting correlations, and the pairing gap systematically diminishes (with no sharp phase transition). Then the exceptional role of pairing in nuclear structure gradually weakens [125]. However, in a large interval of energies above the ground state, pairing determines the transition probabilities and main features of the density of states with a noticeable accumulation above the gap of pushed up levels related to broken pairs, Fig. 2.4.

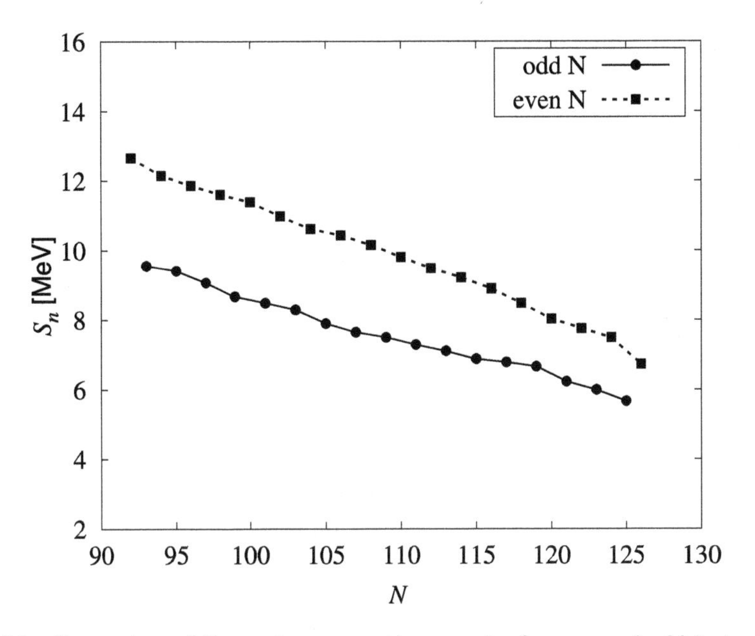

Fig. 2.3. Comparison of the neutron separation energies for even and odd isotopes of mercury ($Z = 80$) shown as a function of the neutron number.

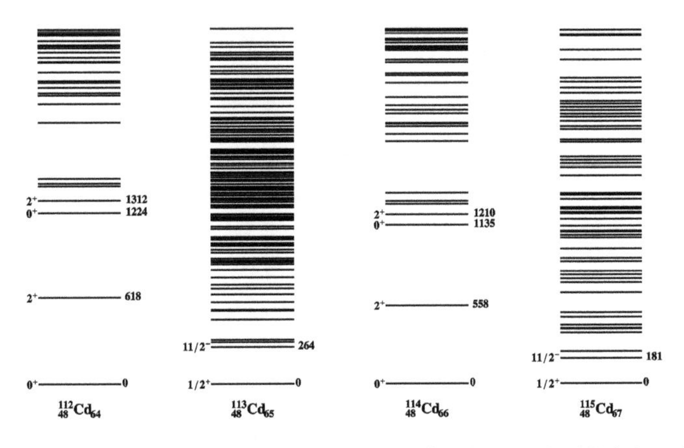

Fig. 2.4. Low-lying energy levels of four isotopes of cadmium, highlighting the gap in the spectrum of even-even nuclides. A few lowest states are labeled with spin and parity and their energies are shown in keV.

Historically, the whole theory of nuclear pairing was developed [30] as a nuclear variation of the BCS theory for macroscopic superconductivity. The preferred formalism was that of the Bogoliubov transformation that entangles the creation and annihilation fermionic operators into a new quasiparticle entity. The elementary excitations coming from breaking a condensate pair are combinations of particle and holes. The energy 2Δ of breaking the pair is essentially the same difference of binding energies that was already shown in Fig. 2.3. Typically, this quantity equals 1.5–2 MeV. In even–even non-magic nuclei, indeed we regularly see an exhausted level density up to excitation energy of this magnitude. Above that threshold, the level density rapidly grows, similarly to what, according to the BCS theory, takes place in superconductors. This corresponds to many two-quasiparticle excitations shifted up through the gap and recovering now after the pair breaking. In difference to macroscopic theory, the standard approximation using the Bogoliubov transformation to new quasiparticles as superpositions of particles and holes is not always good violating the total particle number and giving only an average over several neighboring nuclei, the effect of no significance for macroscopic superconductors. Instead, in nuclei one can use the exact diagonalization with pairing as a representative of the general residual interaction [237] or develop the improved treatment with the exact particle number conservation [244].

The pairing interaction modifies all nuclear observables at not very high excitation energy. We have already mentioned the systematic differences in binding energies of even–even, odd–A, and odd–odd nuclei as the first indicator of pairing correlations. The main part of pairing effects comes from the terms in the two-body Hamiltonian (2.4) corresponding to the scattering of pairs with $J = 0$,

$$H_p = \frac{1}{2} \sum_{12} F_{12} P_1^\dagger P_2, \tag{2.5}$$

where the operators

$$P_1 = \sum_{m_1} a_{j_1 m_1} \tilde{a}_{j_1 m_1} = \sum_{m_1} (-)^{j_1 - m_1} a_{j_1 m_1} a_{j_1 - m_1} \tag{2.6}$$

couple nucleons into $J = 0$ pairs $(1, \tilde{1})$ on time-conjugate single-particle orbitals; the tilde sign marks the time conjugation that is fixed by the phase in Eq. (2.6). In practice, the amplitudes F_{12} frequently are substituted by an effective pairing constant (inside a certain layer around the Fermi surface).

In relatively light nuclei, protons and neutrons occupy the same orbitals and pairs have isospin $T = 1$ with a possibility of all three neutron–proton combinations. In heavier nuclei with $N \neq Z$, the occupied proton and neutron orbitals might be quite different, so that the pairing occurs separately between protons and neutrons. The $T = 0$ proton–neutron pairing is seemingly weaker [239] and corresponds to deuteron-like pairs with angular momentum $J = 1$ while one can argue that the main forces between such pairs prefer their parallel orientation. For example, looking at light odd–odd nuclei, we see that ^{10}B with three deuteron-like pairs on top of the alpha particle has the ground-state quantum numbers $J^{\Pi} = 3^{+}$; the isotope ^{22}Na, also with three $n - p$ valence pairs, in the same way has the ground state 3^{+}; the isotope ^{26}Al with five valence $n - p$ pairs has the ground state 5^{+}, while we know that all even–even nuclei have ground-state spin $J = 0$.

As the number of Cooper pairs in a nucleus is small, the breaking of a pair already considerably diminishes the condensate energy (mainly due to the blocking, Pauli exclusion of certain broken pair states from the interaction volume accessible to other pairs) and in average increases the moment of inertia in the direction of the rigid body value, see the next section. Gradually, the pairing effects, both in deformed and spherical nuclei, are getting less pronounced with the growth of excitation energy. A mesoscopic nature of nuclear superfluidity excludes the sharp phase transition from paired to normal states. Instead, as it was already mentioned, in all nuclear characteristics depending on pairing, we see only smooth crossovers [125] to a non-paired situation where the pairing interaction becomes just one of many possible correlational processes in the nucleus.

The BCS description of nuclear pairing not always adequately reflects the specific underlying physics. Among possible situations, when the characteristic nuclear features, such as pairing, have to be acknowledged in a more detailed (and more appropriate) way than in macroscopic samples, we can mention the near-magic nuclei and nuclei near the onset of deformation. In such applications, it could be physically important to keep track of the exact particle number while the BCS approach, quite appropriate for macroscopic applications, violates this conservation. In such cases, the full diagonalization can be useful, see for example [237], that can correct the fictitious effects of the particle number non-conservation brought in by the pure borrowed BCS approach. The exact solution discovers a noticeable correlation energy lost by the BCS approximation (example of ^{48}Ca in [238]). Another possibility is to generalize the BCS method precisely following the particle number in the amplitudes of the canonical transformation from normal particles to quasiparticles [244]. It can be also mentioned that the complete consideration of pairing effects, for example in a two-dimensional model [10], reveals that the exactly accounted pairing interaction carries also some chaotic effects, however not sufficient to reach the full chaotic stage.

2.6. Collective Rotation

In macroscopic objects consisting of interacting constituents, there are important types of coherent motion not localized on individual particles or specific sites of a crystal lattice. These are usually *waves* of consecutive excitation propagating over a large volume. The quantization introduces corresponding elementary excitations, the renormalized ("dressed") original particles, as well as new bosonic quasiparticles — phonons, magnons, etc. The second-order interaction of electrons with phonons is the main source of low-temperature superconducting correlations in metals leading to effective attraction and pairing. In molecules, we know relatively slow vibrational and rotational degrees of freedom with large energy gaps between them, and fast electronic excitations. In nuclei there is no such a small parameter as $\sqrt{m/M}$ coming from a large mass difference between electrons and ions and leading to these gaps. Nevertheless, complex nuclei demonstrate the existence of qualitatively similar modes of collective motion, certainly with nuclear specifics.

Effects due to pairing modify many important nuclear characteristics, especially in response of the nucleus to typical perturbations (nuclear susceptibilities). It might be technically convenient to consider rotation of a deformed nucleus, that is an intrinsic nuclear excitation [32], as a response to external cranking. This response defines the nuclear moment of inertia and therefore a spectrum of excited states forming rotational bands (based on different intrinsic configurations) with the growing total angular momentum and only slowly changing intrinsic structure.

The collective rotation is a nuclear counterpart to the *Goldstone mode* [106, 218], the gapless branch of the excitation spectrum that appears in quantum field theory as a result of spontaneous violation of continuous symmetries. In the intrinsic frame, the deformation of the nucleus with a fixed orientation in space violates the rotational symmetry that is restored by the possibility of dynamical rotation with rotational bands erected on different intrinsic states and approximately following the classical energy spectrum,

$$E(J) = \frac{\hbar^2 J(J+1)}{2\mathcal{J}}. \tag{2.7}$$

It is known that, in a normal Fermi-system, the corresponding mass parameter $\mathcal{J}$ is, up to small quantum fluctuations, the rigid-body moment of inertia, $\mathcal{J}_{\text{rigid}}$, corresponding, in the classical sense, to the mass distribution for a specific nuclear shape [191] (in the frame rotating together with the deformed nucleus). The nuclear pairing, or as many physicists prefer to formulate, nuclear superfluidity, makes the moment of inertia significantly smaller [30, 265]. However, it is still greater than the hydrodynamic moment of inertia $\mathcal{J}_{\text{rigid}}\beta^2$, where $\beta \approx 0.2 - 0.3$ is the typical nuclear axial deformation parameter (we can mention that some exotic nuclei with the large neutron excess have anomalously large deformation, for example $\beta \approx 0.6$ in the heavy isotope ^{34}Mg [128]). This difference is one of the important consequences of pair correlations; at even stronger pairing the system would behave similarly to a rotating Bose gas of pair molecules with a low hydrodynamic moment of inertia defined only by a relatively small non-spherical crust of the mass distribution. This reminds an intermediate stage of the unfinished *BCS–BEC transition* from the BCS superconductor to the Bose–Einstein condensate of pairs as in a superfluid liquid known in macroscopic physics [104]. As was already mentioned, the formally calculated pair correlation length defined in the BCS theory as $\xi = \hbar v_F/\Delta$, where v_F is the nucleon velocity at the Fermi surface, turns out to be greater than the nuclear size. With the excitation energy growing, there appear specific nuclear phenomena, such as the gradual mixing [26] of the K quantum number (angular momentum projection onto the symmetry axis of the axially deformed nucleus) that is related to the growing density fluctuations.

In the case of a well-developed deformation, the rotational bands with the systematically increasing energy as a function of nuclear spin J can be built on top of any intrinsic configuration formed in the rotating frame. As rotational intervals (2.7) are typically small compared to energies of single-particle excitations, there is an excess of such states at low energies, the so-called *collective enhancement* of the level density, see later. Complementary to specific energy intervals given by Eq. (2.7), there exist certain *intensity rules* for the strongly enhanced multipole, usually quadrupole and magnetic dipole, electromagnetic transitions within the band and weaker interband transitions. Such rules have essentially the geometric nature stressing the collective character of rotational motion that is adiabatic with respect to intrinsic excitations. There are interesting phenomena based on the web of such rotational bands erected on various intrinsic configurations; the bands can cross and mix in the vicinity of crossing [99, 158]. With increase of excitation energy, the moments of inertia of the bands gradually grow approaching on average the rigid body value [191] as a signal of weakening pairing correlations.

In nuclei with ill-defined shapes it might be better to speak about the wave functions of collective variables at a deformation containing various superimposed shapes with different probability amplitudes. Here also there are collective states of rotational type but frequently with not monotonous sequences of nuclear spins. The study of this physics is typically based on the deexcitation of high spin states formed in the heavy ion collisions. As mentioned earlier, this process can start at very high angular momentum, $J \sim 80$, and, by a photon radiation cascade, lead to the existence of traps (irregularities of the spin sequences) along such lengthy bands and appearance of long-lived *isomeric states* which can also show up in a random contact of rotational bands like in the case [19] of the most long lived natural isomer ^{180}Ta.

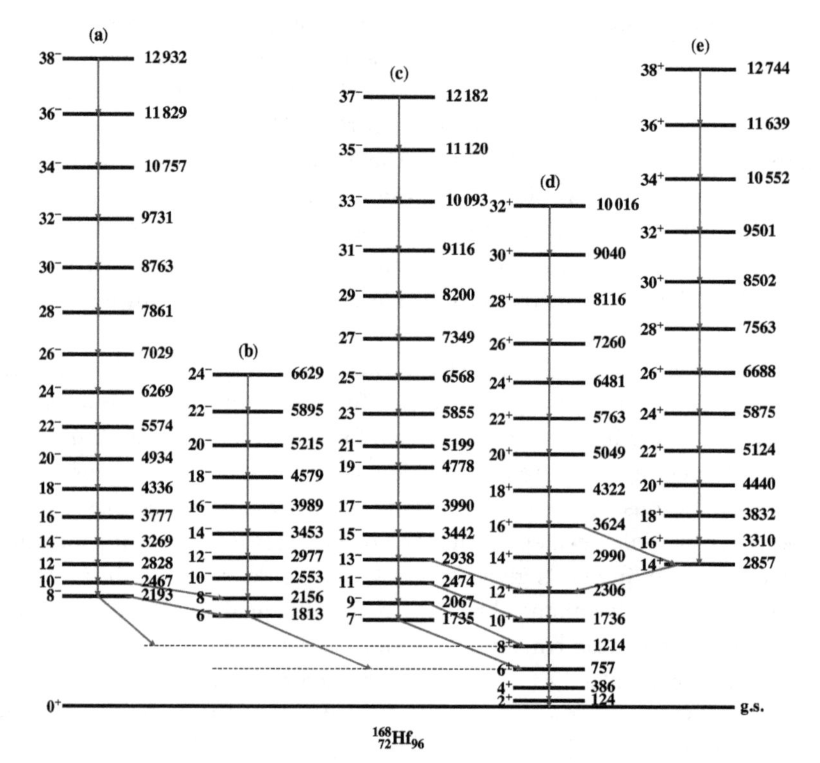

Fig. 2.5. Examples of low-lying rotational bands (a), (b), (c), (d) and (e) in ^{168}Hf. Arrows show interband gamma transitions. See also Ref. [265].

2.7. Vibrational Modes

In molecules, very well-developed *vibrational states* can extend up to a high number of quanta. This is possible due to the large mass difference between electrons and ions separating rotational and vibrational modes. In nuclei, we see embryos of such collective vibrations. The simplest representatives of this branch in the nuclear excitation spectrum are low-lying quadrupole vibrations [31]. The lowest vibrational states $J^{\Pi} = 2^{+}$ are present, with rare exceptions, as the first excitations on top of the 0^{+} ground state in spherical non-magic even-even nuclei. Due to the pairing, such vibrations have low frequency located inside the pairing gap. They can be treated as the slow propagating waves of quadrupole deformation of preexisting pairs. Such modes — low lying *phonons* — are specific for a paired system. In many cases, higher states can be interpreted as multiplets of few-phonon excitations with different couplings of their angular momenta even if the anharmonicity is typically significant.

The collectivity of such excitations leads to strong electromagnetic transitions with excitation and deexcitation of phonons. In spite of the long history of considering such modes as bosonic excitations [31], the interpretation is not always clear, and simple rules which would be valid for harmonic bosonic modes are not exactly fulfilled. With growth of energy, low-lying quasiboson branches of the excitation spectrum usually are disappearing (although there are exceptional cases, like cadmium isotopes with long quasivibrational bands) being immersed in the ocean of states of many-body structure sharing the collective strength.

As the occupancy of upper nucleonic shells grows, the nucleus moves away from the magic numbers and becomes softer as a sign of a growing interaction between valence quasiparticles and resulting growth of the vibrational amplitude being a precursor of static deformation. In some cases, the onset of deformation is rather sharp reminding the macroscopic shape transition, Fig. 2.6. As it is natural to expect, well-deformed nuclei are more rigid. The vibrations of the deformed shape are also observed in many cases but their collective strength is usually lower. The transition between spherical and axially deformed regions can also proceed through intermediate triaxial shapes.

In heavy nuclei, there exist also a trend to the static *octupole* (pear-shaped) deformation. Then, similarly to what is known from molecular physics, the *parity doublets* can appear which are predicted to lead to a significant enhancement of the electric dipole moment (EDM) in an odd-A

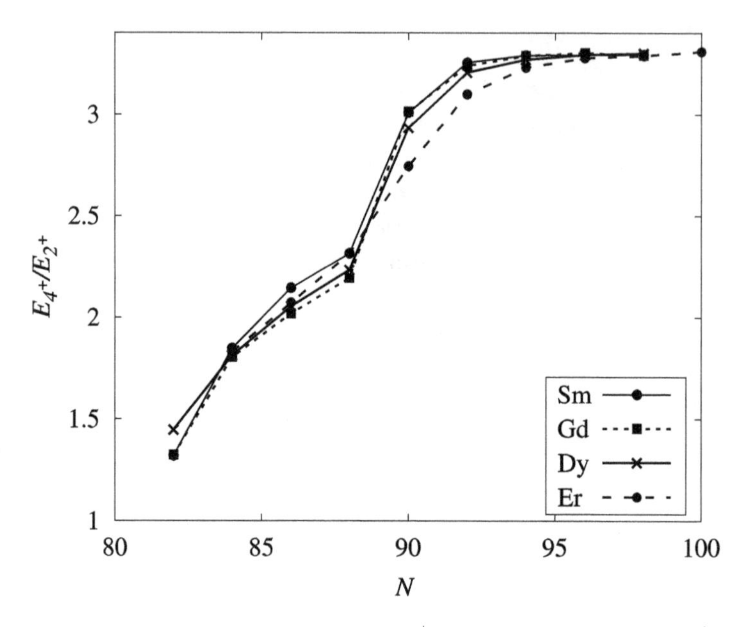

Fig. 2.6. Ratio of excitation energy of the first 4^+ state to that of the lowest 2^+ state for nuclei around $Z = 60$ as a function of neutron number. This ratio shows the transition from the pairing type spectrum ($E_{4^+}/E_{2^+} \approx 1$) to vibrational ($E_{4^+}/E_{2^+} \approx 2$) to rotational ($E_{4^+}/E_{2^+} \approx 3.3$) types.

nucleus [216]. The nuclear EDM, that violates parity and time-reversal symmetry, would be an important signal of physics beyond the Standard Model. In fact, in many soft spherical nuclei there exists a noticeable correlation between quadrupole and octupole vibrational modes [168] that can serve as an EDM amplifier [17, 93], analogously to static deformation.

In addition to low-energy vibrational excitations, there are well-studied collective modes at higher energy, the so-called *giant resonances* of various multipolarities [47, 120]. They are less sensitive to the details of occupancies of the low-lying orbitals being mainly defined by the general elastic properties of the nuclear matter. Such collective excitations are observed as noticeable broad cross-section peaks in various reactions. In the rough approximation, some multipole resonances have main components concentrated close to nuclear surface but also penetrating into nuclear interior. The positions of the main peaks of giant resonances have typical geometric dependences of their centroids as roughly inversely proportional to the nuclear radius, $A^{-1/3}$, while pure surface waves of a liquid drop would have excitation energy $\sim A^{-1/2}$ as follows from the hydrodynamic model.

Giant modes of different symmetry are characterized by the total angular momentum (multipolarity) and parity, J^{Π}, as well as by isospin and specific contributions of spin degrees of freedom. The Gamow–Teller (GT) resonances are excited by the same spin–isospin charge exchange reactions in a neighboring nucleus as in the GT beta decay.

The first and best studied excitation of this type is the *giant dipole resonance* (GDR) with the quantum numbers $J^{\Pi} = 1^{-}$ and a rough energy centroid dependence $\sim 80\,A^{-1/3}$ MeV, Fig. 2.7. With the nuclear center of mass at rest, this excitation is essentially the isovector dipole oscillation of protons with respect to neutrons and as such can be estimated with the help of the empirically known nuclear symmetry energy [265]. This interpretation is supported by examples of splitting of the GDR in well-deformed nuclei where the dipole vibrations become possible along different axes and, correspondingly, with different frequencies, Fig. 2.7. In a microscopic consideration, the parent wave function of the GDR (and similarly for other giant resonances) is a coherent superposition of many particle-hole excitations from the ground state, in this case to the next shell of opposite parity. This mode is moved up in energy compared to the pure shell-model excitation by the residual interactions responsible for the symmetry energy. According to the *Brink–Axel hypothesis* [22], such a resonance, due to its

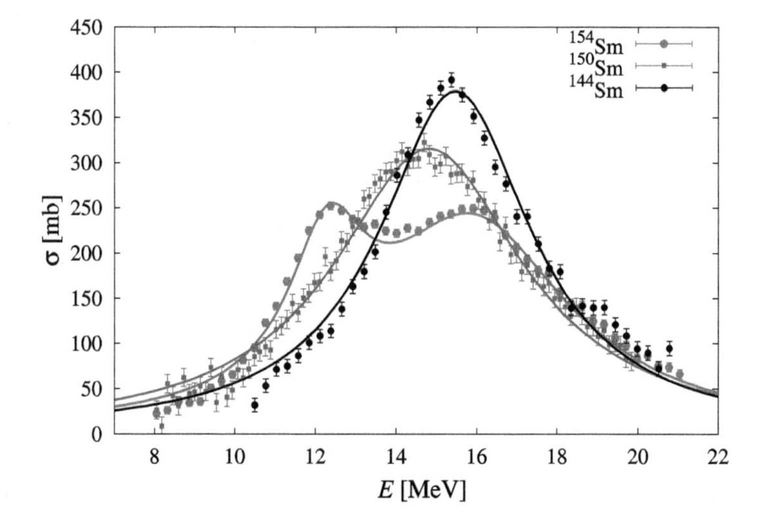

Fig. 2.7. GDR in isotopes of samarium. The splitting of the GDR as we transition from the spherical isotope ^{144}Sm ($N = 82$) to the deformed one, ^{154}Sm ($N = 92$) is observed; experimental data [56] fitted with Lorentzian curves.

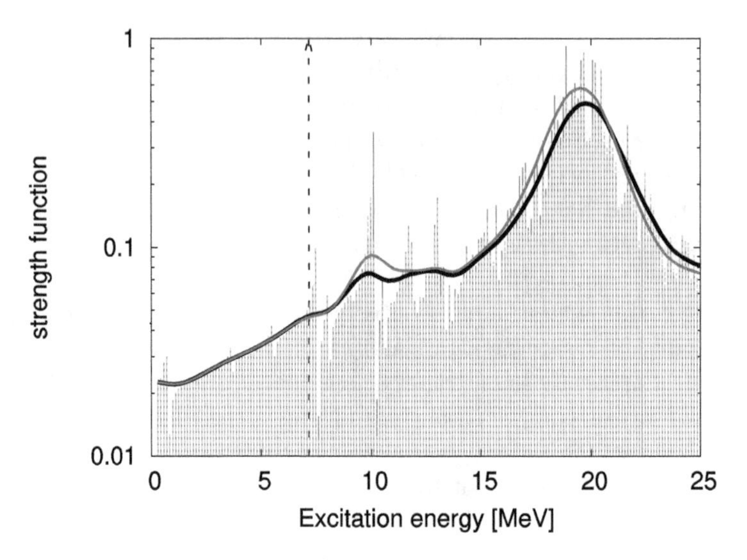

Fig. 2.8. Isovector dipole strength in ^{22}O. Calculations with continuum show an emergence of a pigmy resonance. The neutron decay threshold is indicated with a vertical dashed line. Coupling to the continuum of weakly bound neutrons, considered in the calculations shown by red line, results in the formation of a pygmy resonance. See Ref. [243] for details.

fundamental physical nature, can be excited on a background of practically any additional nuclear excitation. The GDR concentrates the large part of the total isovector dipole strength; the remaining strength, less collective, is located close to the shell-model position of the simple multipole excitations and forms here the so-called *pigmy resonances* [21], Fig. 2.8.

The giant resonances are not stationary nuclear states as it is immediately seen by their large energy widths. First of all, typically a considerable fraction of the giant resonance peak is already in the continuum part of the energy spectrum and undergoes irreversible decay. The corresponding decay width is usually called $\Gamma^{\uparrow}$. But frequently the main component of the total width is $\Gamma^{\downarrow}$ that gives an inverse time of decay of a coherent motion into internal states of complex configurations, the microscopic analog of mechanical *damping*, or friction. The actual lifetime of the collective giant resonance may correspond just to few oscillatory periods. The underlying states, especially for the resonances shifted by residual interactions up in energy, are nuclear configurations which are more complex than those of just a particle–hole nature. They, in turn, may have their own widths indicating their finite lifetimes if they are located above the continuum threshold. The interrelation between giant collective modes, their complex background, and corresponding experimental signals depends on reactions used in the experiment (*nuclear kinetics* [46, 213], see the discussion in Secs. 6.5 and 6.6). If such reactions first populate background states at high enough energy, they may decay back to continuum before having a chance to excite a collective mode. This can lead to the visible loss of collectivity [97, 251].

2.8. Mechanism of Collective Motion

Among various residual interactions, we have to specially discuss those which coherently involve many single-particle degrees of freedom. The corresponding quantum states typically reveal their nature in observed strong probabilities of excitation and deexcitation. Of course, the full diagonalization of the Hamiltonian matrix in a sufficiently large orbital space would find all stationary states, including those we call *collective*. Here we show instead just a skeleton for qualitative understanding the process of formation of a collective mode.

The characteristic feature of a collective state is a large matrix element of a corresponding operator (frequently of obvious classical meaning) between this state and the ground or one of lower lying states. This is what is usually seen through a high radiation probability. In the opposite direction, this state can be earmarked by a strong excitation from the ground state driven by a coherent superposition of simple operators, typically of particle–hole (or quasiparticle in the pairing scheme) type. Appropriate examples are the low-lying vibrations of quadrupole or octupole symmetry, as well as relatively high-lying dipole oscillations giving rise to a giant resonance. A big difference between collective modes of these two types is in the fact that low-lying vibrations are, with a high precision, actual stationary states of the many-body nucleus, while the giant resonances represent oscillations of a certain multipolarity, reminding the similar waves in a liquid drop, which are *damping* quasistationary states gradually dying away into complicated background states, an analog of macroscopic friction. The low-lying vibrational states decay only by electromagnetic radiation while the giant resonances at higher energy have large widths that may include also real particle decays into continuum.

The mechanism of formation of a collective state can be illustrated by the well-known schematic consideration. Let $|k\rangle$ be mutually orthogonal quantum states of "simple" nature with the fixed exact quantum characteristics of spin and parity, for example of particle–hole or quasi-particle nature. Their unperturbed energies are the eigenvalues ϵ_k of the Hamiltonian H_0 that does not include the part H_q of the total Hamiltonian responsible for the creation of a collective state. The mode of interest is formed by the excitations driven by simple operators q_k of given symmetry, for example of the same multipole and isospin type. Including the interaction between the basis states through this collective mode we come to the simplified *multipole–multipole* Hamiltonian with the matrix elements

$$H = H_0 + H_q, \quad H_{kk'} = \epsilon_k \delta_{kk'} + \kappa q_k^* q_{k'}. \tag{2.8}$$

Here the real parameter κ regulates the collective strength of this multipole mode. The stationary states $|\alpha\rangle$ of this schematic Hamiltonian are superpositions of original decoupled states $|k\rangle$,

$$|\alpha\rangle = \sum_k C_k^\alpha |k\rangle. \tag{2.9}$$

For the factorized interaction (2.8), the solution of the Schrödinger equation,

$$H|\alpha\rangle = E^\alpha|\alpha\rangle, \tag{2.10}$$

is obvious,

$$C_k^\alpha = \frac{q_k^*}{E^\alpha - \epsilon_k} \kappa \sum_{k'} C_{k'}^\alpha q_{k'}. \tag{2.11}$$

Taking a non-trivial solution when the vector of amplitudes C_k is not orthogonal to the multipoles q_k, we come to the secular equation defining the energies E^α of eigenstates,

$$\frac{1}{\kappa} = \sum_k \frac{|q_k|^2}{E^\alpha - \epsilon_k}. \tag{2.12}$$

In the absence of the multipole interaction, $\kappa \to 0$, we return to the original energies ϵ_k as the poles of the right-hand side in Eq. (2.12). As seen from the simple graphic solution of the secular equation, Fig. 2.9, all roots E^α are in between the unperturbed energies ϵ_k. The exception is the edge state that is shifted to the left for $\kappa < 0$ (attraction) and to the right for $\kappa > 0$ (repulsion). For sufficiently large $|\kappa|$, this state is repelled far away from the quasicontinuum of other only slightly perturbed roots. In such a situation, all amplitudes C_k^α for a given root α are of the same sign which is a signal of coherent behavior. The emerging collective excitation,

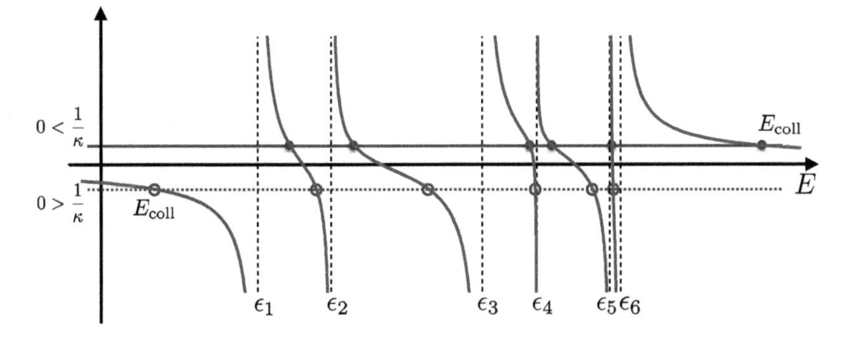

Fig. 2.9. Graphical solution of Eq. (2.12) where the lines on the levels $\pm 1/\kappa$ (shown in red) are crossed with the graph of the right-hand side (shown in blue) that has multiple poles at $E = \epsilon_k$. The roots E_α correspond to the crossing points.

being an analog to the sound in an elastic medium, is what we can call a *phonon*.

In the standard shell model based on the oscillator scheme, the parity of the shells alternates. Due to the spin–orbit coupling, a single-particle level with higher angular momentum $j = \ell + 1/2$ moves down and frequently turns out to live in the previous shell, an *intruder* [105, 265]. This acts in the direction of equilibrating the appearance of two parities in the spectrum. Nevertheless, the lowest collective phonon is of the quadrupole type, $J^\Pi = 2^+$. Such a shape oscillation (the solution with $\kappa < 0$) mainly takes place close to the nuclear surface but still penetrates the interior as well. As mentioned above, the energy of this state as a distortion of spherical pairs is usually located inside the pairing gap, so low in energy that even its overtones, the two-phonon triplet, $J = 2, 4, 0$, can appear in many nuclei (with some anharmonicity). The higher overtones are rare, usually they are already immersed in the sea of enhanced level density above the pairing threshold of 2Δ. When the quadrupole phonon becomes too soft, the vibrating nucleus spends a significant fraction of time with a deformed shape, and the effective vibrational amplitude grows as a hint of approaching the phase transition to static deformation.

In the case of $\kappa > 0$, the phonon frequencies are pushed up, higher than the place of the original simple particle-hole (or pair breaking in the presence of pairing) excitations of the same symmetry. The GDR with the parity change is the most pronounced excitation of this type. It is not sensitive to the details of low-lying dynamics being defined [157] essentially by symmetry energy that reacts to the neutron–proton matter displacement in the dipole excitation in the center-of-mass frame. Therefore, the above-mentioned Brink–Axel hypothesis [22] of the GDR erected on top of an arbitrary intrinsic state can make sense. As it was also mentioned, the splitting of the GDR gives a signal of static deformation, while the coupling between the GDR and many-body underlying states of the same symmetry provides the mechanism of damping that determines the width of the GDR and its effective lifetime. The interaction between the GDR and its environment usually happens above threshold for evaporation of particles [213], and, even higher in the continuum, in some reactions it was possible to see the overtone, double GDR [148].

A special type of collective correlations is presented by the *cluster formation*. The trends to clusterization are clearly pronounced in several regions of the nuclear chart. Light nuclei are known by the high probability of alpha-type correlations. The beryllium isotope ${}^{8}_{4}\mathrm{Be}_4$ decays into two alphas being essentially their long-lived bound state of molecular type, the lifetime is 0.7×10^{-16} sec. In the opposite direction, the capture of the additional alpha-particle with the occupation of the famous Hoyle state in ${}^{12}\mathrm{C}$ is an important step in the stellar nucleosynthesis (*triple alpha reaction*). The next nucleus of this line, ${}^{16}\mathrm{O}$, was observed [76] to decay from the excited state through two ${}^{8}\mathrm{Be}$ and then into four alphas after electroexcitation in the electron storage ring with the superthin nuclear jet target. This trend can be traced further [24] to nuclei as ${}^{24}\mathrm{Mg}$, ${}^{28}\mathrm{Si}$, ${}^{32}\mathrm{S}$, and maybe ${}^{56}\mathrm{Ni}$; the data hint on the presence of the wave function component similar to an alpha-crystal.

The formation of alpha-type clusters most probably continues in the region close to the last $N = Z$ magic nucleus, ${}^{100}\mathrm{Sn}$. Here one observes alpha-decays which, after excluding the barrier penetration factor, turn out to be the strongest alpha-decays in the entire nuclear chart [149]. Such alpha-correlations can be microscopically considered in the same spirit as pairing [201] but this theory still is not sufficiently developed. Of course, heavy cluster decays and long chains of alpha-decays are well known for heavy or superheavy nuclei [185], and nuclear fission is the limiting case of this gallery of nuclear transmutations.

2.9. Symmetries

Nuclear stationary states are naturally divided into classes characterized by experimentally established global quantum numbers. Energy and total angular momentum (*nuclear spin*) with its projection on the quantization axis are exactly conserved; isospin — approximately, perturbed primarily by the Coulomb interaction between the nucleons. It would be natural to expect the parity violation effects by the weak interactions to be very small but, as will be discussed later in more detail, there are mesoscopic effects enhancing such violations under some conditions, as it is clearly seen in special experiments.

The total Hilbert space of a nucleus is divided into blocks with certain values of general constants of motion. The blocks are dynamically driven by the common nuclear Hamiltonian. In many cases, the properties of different blocks are mutually correlated. This is especially clear in the case of rotational bands where the total spin grows keeping the neighboring band members of different spin J with similar intrinsic structure seen in regular geometric energy intervals and transition probabilities (for example of collective E2 radiation) between the band members. As mentioned earlier, in soft nuclei without a well-deformed shape, very long rotational bands excited at high angular momentum in a heavy ion collision, may have many deviations from the regular spin sequences; this leads to traps and isomeric (long-lived) states in the sequence of radiative transitions down the band.

Some properties are only weakly determined by the specific features of nuclear dynamics in individual nuclei. Instead, the main role here belongs to the symmetry of the orbital space, see Secs 2.10 and 2.11. This can be studied substituting the realistic nuclear Hamiltonian either by a *random matrix* with the same global constants of motion determined by symmetries, or even by a dynamics of several interacting bosons [127] that mimic fermionic pairs. One of the significant statistical results is the fact of the prevalent appearance of states with total spin $J = 0$ as ground states in systems with a *randomly* picked Hamiltonians within a simple (even a single j-level) orbital space for an even number of nucleons, Fig. 2.10. This happens in spite of the fact that the dimension of the class of states $J = 0$ is noticeably lower than for some other values of J [169, 261]. Most probably, the statistical width of the representation with the lowest symmetry is greater which is confirmed by the fact that the probability of the ground state with the maximum possible for the given orbital space spin $J_{\max}$ is also enhanced while the wave function of such a state is frequently unique and therefore more susceptible to large deviations not spoiled by the disordered interference of many contributions.

The important aspect of the nuclear symmetry problem is the manifestation of fundamental global symmetries and their violation. Weak interactions in nuclei are revealed mainly in the beta decay and related processes.

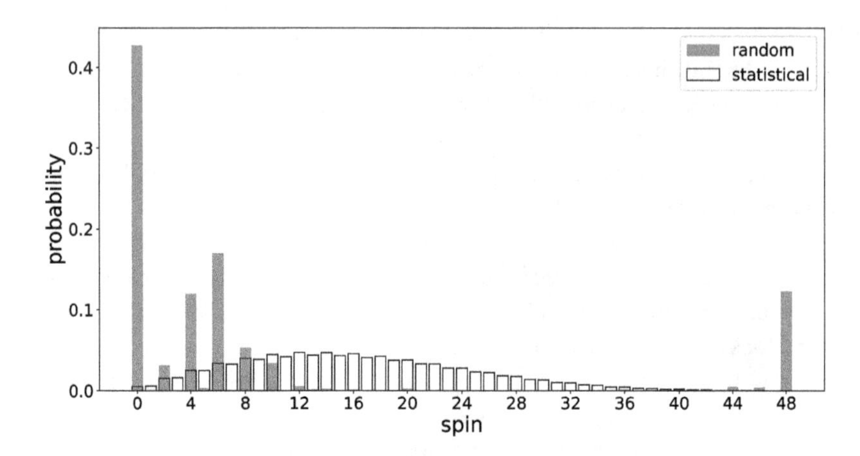

Fig. 2.10. Systematics of ground state spins for a system of eight fermions on a single degenerate $j = 19/2$ level. Reduced matrix elements of the two-body interaction are selected at random from the normal distribution, so that in the case of two particles the Hamiltonian is that of the Gaussian Orthogonal Ensemble. The results from this two-body random ensemble are compared with the statistical expectation determined directly by the fraction of states with a given spin available in the system.

The left-handed character of weak currents is seen in violation of symmetry with respect to spatial inversion, $\mathcal{P}$, and charge conjugation symmetry, $\mathcal{C}$: neutrino in the beta decay are left-polarized and antineutrino right-polarized. The combined $\mathcal{CP}$ inversion is approximately kept as a good symmetry of nuclear weak processes. We still do not know if the neutrino and antineutrino are Dirac particles with their own antiparticles or they are essentially the same Majorana fermions. Masses of the mixed neutrinos of different generations are very small but still unknown. All such questions are extremely important for the current standard model of elementary particles and great cosmological problems of the universe, as well as for the understanding the stellar nucleosynthesis and resulting abundances of chemical elements. Here we can stress that the mesoscopic character of the nucleus can help in the studies of global symmetries. The nuclear mechanisms (to be discussed later in more detail) lead to the enhancement of violation of spatial parity by orders of magnitude.

The existence of the electric dipole moment (EDM) in a stationary state of an isolated atom, molecule, nucleus, or an elementary particle would violate together $\mathcal{P}$ symmetry and invariance under time reversal, $\mathcal{T}$. Indeed, for a stationary state $|JM;\alpha\rangle$ of an individual quantum state with angular momentum J, its projection M onto the quantization axis, and other quantum numbers α, the diagonal with respect to J and α matrix element of a vector operator $\mathbf{v}$ is proportional to the similar matrix element of the angular momentum operator as the only strictly conserved vector in the rest frame,

$$\langle JM\alpha|\mathbf{v}|JM'\alpha\rangle = \xi\langle JM\alpha|\mathbf{J}|JM'\alpha\rangle. \tag{2.13}$$

This is a very old vector model still valid in quantum theory. The proportionality constant ξ is specific for every operator and corresponds to

$$\xi = \frac{\langle(\mathbf{v}\cdot\mathbf{J})\rangle}{J(J+1)}. \tag{2.14}$$

If the state under study has certain parity, while the operator $\mathbf{v}$ corresponds to a polar vector, the mean value vanishes as the parity has to be changed in this matrix element. An additional restriction comes from the time reversal symmetry. This operation changes M to $-M$ as the angular momentum changes sign. This means that the operator $\mathbf{v}$ also has to change sign under time reversal. The electric dipole moment operator is a polar time-even vector, therefore it is forbidden by both symmetries. In contrast to this, the magnetic moment is completely allowed being an axial time-odd operator similar to the angular momentum. An intermediate situation takes place for the so-called *anapole moment* constructed as a polar time-odd operator of the type $[\mathbf{r}\times\mathbf{s}]$ forbidden only by parity conservation.

The current Standard Model predicts the violation of the $\mathcal{PT}$ invariance on a very low level, orders of magnitude below experimental possibilities. As an example we can mention the last measurement [1] of the neutron EDM with the result $(0.0\pm1.1)\,10^{-26}\mathrm{e}\cdot\mathrm{cm}$ while the Standard Model prediction is on the level of $10^{-31}\,\mathrm{e}\cdot\mathrm{cm}$. The search for such extremely small effects of physics beyond the Standard Model can be facilitated by nuclear collective effects, namely the combination of quadrupole and octupole collectivity [216]. As will be explained later, the statistical properties of spectra in principle can distinguish systems invariant under time reversal from those where such invariance is absent. All such phenomena stress that the mesoscopic structure is capable to couple the deepest microscopic physics to possible macroscopic manifestations.

2.10. Transition to Complexity

Traditionally, the low-energy physics of microscopic fermionic systems can be discussed in terms of Fermi-liquid theory started by Landau [173]. For complex nuclei, this approach was reformulated and practically used [157] by Migdal and his school, including, among other typically nuclear features, pairing correlations. Usually, such theories consider single-particle excitations — quasiparticles — not far from the Fermi surface. Their interactions create global nuclear features, starting with the mean field, as well as collective motion. As excitation energy grows, the density of nuclear states increases exponentially just because of combinatorics [85, 267]. As a consequence, the interactions between quasiparticles and collective waves become effective creating the high density of stationary nuclear states as very complicated superpositions of simple modes.

We can remain on the grounds of Fermi-liquid propagating this approach to higher excitation energy and allowing the strong mixing similar to the growth of temperature inside the nucleus as a closed quantum system. We are still considering a closed mesoscopic system with no external thermostat (heat bath). After an initial excitation, the system is evolving to the intrinsic equilibrium distribution of available conserved energy over many degrees of freedom, collective and non-collective. Many interactive events including the wave damping and collision-like incoherent scattering of quasiparticles contribute to this process of *internal thermalization* and mixing of simple shell-model partitions. As a result, the observable characteristics of energetically close states with the same global constants of motion become very similar to what is expected for the equilibrium microcanonical ensemble. The structure of eigenstates changes smoothly with excitation energy as it is supposed for *thermodynamic* quantities in equilibrium.

In real life, the excited states emerge as a result of some external process, so that the time factor enters the game as the equilibration of a state initially created in a nuclear reaction has to be established earlier than the excited system decays back into continuum. This is almost certainly the case if the process is driven by the electromagnetic forces and is finished with photon radiation. At energy higher than the particle decay threshold, the lifetime can be shorter than the equilibration time. The difference with macroscopic equilibrated systems again is in the opportunity to work with individual quantum states including resonances. Neglecting the probability of decay, one can still theoretically consider the excited states as stationary members of a statistical ensemble.

In the full equilibrium, the system is expected to have universal properties determined by the particle number, excitation energy, and the set of exact quantum numbers. The simple quantum-mechanical level repulsion builds some kind of a ladder of states with locally uniform and slowly changing along excitation energy consecutive spacings. The global observables for close states are also smoothly changing along the ladder becoming therefore also thermodynamic variables, see below Fig. 5.16. The details of the close wave functions, normalized and mutually orthogonal, are subject to fluctuations which are similar to the normal thermodynamic behavior but without an external heat bath.

At this stage it is possible to turn to a complementary description based on the ideas of quantum chaos [51, 186]. The statistical distributions of typical spectral characteristics turn out to be similar to the predictions of *random matrix ensembles* [155]. Several review papers, for example [248, 260, 264], make the thorough study of the exact nuclear shell-model results (no chaotic elements in the solution for the fixed nuclear Hamiltonian in a restricted orbital space) juxtaposed to the predictions of random matrix ensembles, first of all the Gaussian orthogonal ensemble (GOE) based on the unitary and time-reversal invariant probabilistic dynamics. Similar results are obtained for complex atoms [90] and for the two-body random ensembles [92, 137, 138] where, in contrast to the GOE, the interaction consists of random matrix elements for two-body collisions only, with greater similarity to the realistic shell-model versions. In such ensembles, the matrix in fact is not completely random as the same matrix element for a two-body process appears repeatedly on various backgrounds of other particles (spectators); nevertheless, one can see similar manifestations of chaotic behavior.

2.11. On the Road to Chaotic Dynamics

With increase of excitation energy, already the simple mean-field picture predicts the exponential growth of the level density due to combinatorics of single-particle excitations [85]. When the energy spacings s between the simple states with the same spin-parity quantum numbers go down, the wave functions become sensitive to all types of possible perturbations. As we have already stated, the coherent interactions like multipole-multipole forces create collective superpositions giving rise to low-lying vibrations or to high-lying giant resonances. But there are remaining "non-collective" combinations of simple states influenced by all other interactions, like incoherent collision-like processes. This trend can be also strengthened by weakening of pairing above the threshold of 2Δ and corresponding accumulation of broken pairs. The result is a strong mixing of simple mean-field states in parallel to the growth of energy.

Inside the class of states with the same global quantum numbers, the motion to greater excitation energy brings up the states which underwent more and more acts of mixing. In Fig. 2.11 of energy levels as a function of the artificially increasing mean interaction strength, we see the evolving dense web of levels while their wave functions become more and more complicated (with respect to the same initial basis) due to incoming admixtures. In any close "collision" of two states, even weak residual interactions V become effectively strong leading to the level repulsion according to the simple two-state problem,

$$(\epsilon_1, \epsilon_2) \Rightarrow E_{\pm} = \frac{1}{2}(\epsilon_1 + \epsilon_2 \pm \sqrt{(\epsilon_1 - \epsilon_2)^2 + 4|V|^2}). \tag{2.15}$$

In these processes, all crossings are *avoided* as any real degeneracy would require the simultaneous disappearance of the unperturbed energy difference and the mixing matrix element that is highly improbable. As a result of such evolution, we should come to a *disordered crystal* of levels which, after any mixing encounters, have quite similar (still orthogonal!) wave functions and close energy. Here we deliberately put aside that, in a real nucleus, some of these states, or all of them, would already become resonances in continuum with a finite lifetime τ or corresponding complex energy $E - i\Gamma/2$ where the width Γ is $\hbar/\tau$. In practical applications of this philosophy, one can have in mind that there exists a sufficiently wide region of spectrum above thresholds where the widths are still small, and the resonances do not overlap. The openness of the system will be a practically important subject of our additional consideration later.

The giant resonances are, from this viewpoint, just local accumulations, in a spectral interval, of special collective signatures, namely a certain multipole strength and related quantum characteristics. They share their collective strength with the sea of underlying very complicated, really stationary (or quasistationary in the continuum) states located around this specific energy. Sometimes this is called "*scars*" over the ocean of typical underlying states [131]. If it is possible to trace the genealogy of the quantum problem from its classical analog, the scars are concentrations of the quantum probability along unstable classical periodic orbits. The corresponding damping width is what we have mentioned as $\Gamma^{\downarrow}$ in contrast to the width $\Gamma^{\uparrow}$ [89] due to the real decay into continuum.

The evolution process described above can be called *chaotization*, or *stochastization*. Its main trends are universal for any system of interacting quantum constituents but specific features depend on the mean interaction

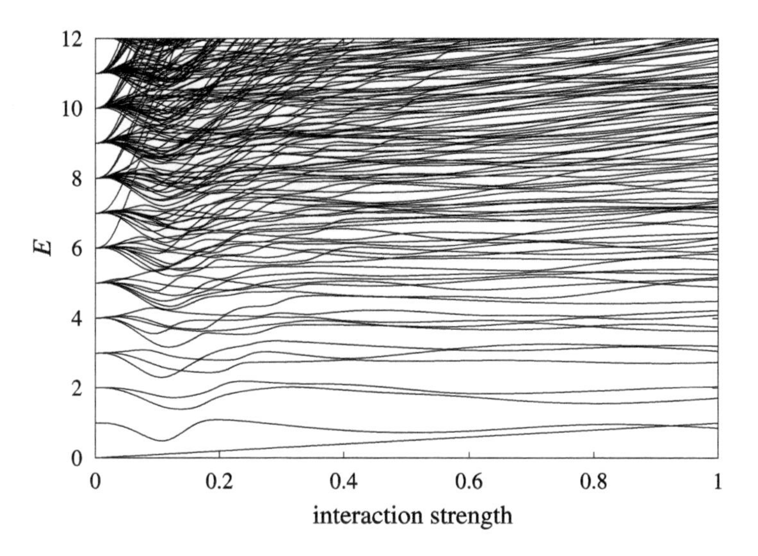

Fig. 2.11. Typical evolution of energy levels in a many-body system as a function of the mean interaction strength. All levels have the same exactly conserved quantum numbers.

strength, systematics of original quasiparticle levels, and symmetry of the mean field, as well as on the proximity of the continuum. We conclude that there are two complementary views of this process and its resulting picture. We come to the quantum chaotic structure that can be considered as a realistic approximation to the limiting mathematical construction described by *random matrices*. The class of those matrices is determined only by the most general symmetry features of the system. The alternative picture is that of statistical mechanics as the result can be discussed in thermodynamical terms of entropy and temperature in spite of the absence of any external heat bath. Both pictures are useful and complement each other. Both of them are quite general and in this way can provide a relatively new approach to the old problem of statistical equilibrium and thermalization.

Chapter 3

Many-Body Nuclear Complexity

3.1. Nuclear Shell Model as an Instrument

To study theoretically single-particle, collective, and statistical properties of nuclei, especially at different excitation energies, we need a reliable computational mechanism. The desirable precision depends on the subject of interest. Near the ground state, physicists are interested in exact properties measured by the experiment: level energies and quantum numbers of individual states, transition probabilities, and mean values of observables, such as electromagnetic moments, density and charge form-factors, etc. This is achieved in various versions of the nuclear shell model currently united under the banner of *configuration interaction.*

We will be mainly relying on the shell model approaches which keep track of exact quantum numbers of many-body quantum states [53, 55, 247, 260]. Therefore, we use the variants with exact conservation of angular momentum and parity; weak interactions violating parity can be included as small perturbations. Often we also assume the isospin conservation; its violation can be easily accounted for. Typically we will not start *ab initio* as it is not our purpose to study meson physics, or even deeper — starting with quarks and gluons. Instead we introduce a generic effective Hamiltonian that includes a limited number of spherical mean field orbitals with quantum numbers appropriate for a certain region of the nuclear chart, and effective interactions, usually of two-body type although many-body components can be always added. In practical applications, the parameters of this effective Hamiltonian are fit by reliable experimental data, mainly in the region near the ground and low-lying states, and extrapolated to neighboring nuclei and higher energies. In this way we come to the popular versions of the shell model which certainly provide a reasonable approximation to more fundamental future theories.

There are traditional criteria of the quality of the shell models, first of all correct properties of low-lying states, their energy spectrum, symmetry, and agreement with qualitative features of experimental results. It is necessary to provide a good level density up to the excitation energy where the practically unavoidable truncation of the orbital space becomes visible as important higher single-particle orbitals are absent in the model. Then we have a double use of the whole construction: up to some energy — direct description and reliable explanation of experimental data, and at higher energy — study of statistical properties of the reasonable physical model for a many-body quantum system. Here the theoretical results can be juxtaposed to the data from specific reactions which would require a modification of the shell model by explicit introduction of continuum states. This will be illustrated in the following chapters.

Some very reliable shell model versions, such as for example the widely used USD interactions [55, 260] for the nuclei from oxygen-16 to calcium-40, exist for light and moderately heavy nuclei. The number of valence nucleons in these nuclei is already large enough to see the evolution of the mean field, single-particle and collective features, as well as the advent of complexity and quantum chaos. The dimensions of Hamiltonian matrices are in the limits of several thousand (for a sector of states with the fixed total angular momentum values) so the exact solution with numerical diagonalization is usually possible. The lower part of the obtained spectrum, up to approximately 12–15 MeV excitation energy can be directly compared with data, both for characteristics of individual levels and transitions and for the behavior of the level density. The high-energy part of the shell model variants with only the discrete spectrum of quantum states should be interpreted as a useful model of a quantum mesoscopic system, while its direct comparison with experimental data would require explicit increase of the orbital space and correct treatment of the continuum.

3.2. Specific Example of the Shell Model

We illustrate the formulation of the shell model in a restricted orbital space by the long studied and widely used *sd*-model. Here we have only three valence orbitals explicitly included in the diagonalization: $d_{5/2}$, $s_{1/2}$, and $d_{3/2}$; they are mentioned in the natural order of growing energy coming from the spin–orbit splitting of the *d*-levels. This requires first three parameters, mean-field positions ϵ_j of these three orbitals, which can regularly change from nucleus to nucleus reflecting, among other things, a gradual evolution of nuclear size. The single-particle wave functions are taken as those of a three-dimensional isotropic oscillator which facilitates calculations but may create problems of a wrong asymptotic behavior later when the continuum effects and nuclear reaction physics enter the game. The neutron and proton energies are taken the same that, certainly, can be adjusted, especially for the nuclei away from the line of stability. In the secondary quantization, this part of the total Hamiltonian is

$$\hat{H}_0 = \sum_1 \epsilon_1 \hat{a}_1^\dagger \hat{a}_1, \tag{3.1}$$

It is convenient to use in general formulations a unified numerical index, like 1 in this equation, that combines all single-particle quantum numbers in a spherical potential field: isospin t_3, radial quantum number, angular momentum j, orbital momentum ℓ (and therefore orbital parity $(-)^\ell$), and projection $j_z = m$. Single-particle energies ϵ_j (independent of m) are parameters of the model that can slightly vary from one isotope to another.

The main part of the model is the residual interaction. It is taken as a set of all allowed two-body collisions with total angular momentum and isospin of the interacting pair conserved. In the *sd*-orbital space there are 63 independent two-body matrix elements. Here we have in mind the so-called *reduced matrix elements* which are the scalar quantities obtained from the full matrix elements by separation of the universal Clebsch–Gordan symbols carrying the dependence on magnetic quantum numbers and vector coupling for a colliding pair. The reduced matrix elements are $V_{j_1 j_2; j_3 j_4; JT}$ which are antisymmetrized with respect to two initial (3, 4) and two final (1, 2) orbital states. Assuming the invariance with respect to time reversal, all matrix elements are real. The angular momentum J and isospin T of the pair are preserved in the interaction event described by the given matrix element. The corresponding numbers typically accepted in the *sd*-model are given below in Table 3.1. Not all of them are completely fixed by the experiment, a significant part can be varied in certain limits; the interested reader can modify this input and find the corresponding changes of the results. On top of this intrinsic uncertainty, we have a freedom of arbitrarily scaling individual matrix elements in order to study and clarify their role in the whole process. The possible addition of many-body forces will even accelerate the appearance of chaotic features.

We have to stress that there are no randomness in this formulation if used for the description of realistic spectra when the orbital energies and interaction matrix elements are defined by the agreement with the data. But these parameters can be also taken as random numbers generated by some statistical distribution (and later we will use this approach along

Table 3.1. Effective interaction matrix elements $V_{j_1 j_2; j_3 j_4; JT}$ in the sd-space for the USDB model [55]. The matrix elements are given in the units of MeV; numbers 1, 2, and 3 in the first four columns correspond to $0d_{3/2}$, $0d_{5/2}$, and $1s_{1/2}$, respectively; the following columns show values of J and columns with matrix elements for $T = 0$ and $T = 1$ as labeled.

j_1	j_2	j_3	j_4	J	$T = 0$	J	$T = 1$
1	1	1	1	3	−2.966	2	−0.0974
1	1	1	1	1	−1.6582	0	−1.8992
1	1	1	3	1	−0.8493	2	0.3494
1	1	3	3	1	0.1574	0	−1.0150
1	3	1	3	2	−1.8504	2	−0.3034
1	3	1	3	1	−4.0460	1	0.5158
1	3	3	3	1	−0.9201		
2	1	1	1	3	1.4300	2	−0.5032
2	1	1	1	1	0.1922		
2	1	1	3	2	−0.4429	2	0.3713
2	1	1	3	1	1.6231	1	−0.0456
2	1	2	1	4	−4.6189	4	−1.4447
2	1	2	1	3	−1.2124	3	0.7673
2	1	2	1	2	−4.2117	2	−0.1545
2	1	2	1	1	−6.0099	1	0.6556
2	1	2	3	3	1.2526	3	−0.5525
2	1	2	3	2	−0.6464	2	−0.3147
2	1	3	3	1	2.0226		
2	2	1	1	3	1.1792	2	−1.2187
2	2	1	1	1	1.6647	0	−3.1025
2	2	1	3	1	0.0272	2	0.8866
2	2	2	1	3	2.3102	4	−1.3349
2	2	2	1	1	3.4987	2	−0.2137
2	2	2	2	5	−4.3205	4	−0.2069
2	2	2	2	3	−1.6651	2	−1.0007
2	2	2	2	1	−1.3796	0	−2.5598
2	2	2	3	3	−1.2167	2	−0.9317
2	2	3	3	1	−0.5344	0	−1.5602
2	3	1	1	3	0.0968	2	−0.3173
2	3	1	3	2	−2.5110	2	1.6131
2	3	2	3	3	−4.1823	3	0.6841
2	3	2	3	2	−0.3154	2	−0.9405
3	3	3	3	1	−3.7093	0	−1.6913

with the realistic sets of matrix elements as in Table 3.1). In any case, the geometric properties and conservation laws should be preserved while the numerical values of elements can be randomized. If we imagine a matrix of the full Hamiltonian in N-body space, our limitation by two-body scattering makes this matrix sparce and remote from chaoticity as any two-body process in a many-body system can happen on many different backgrounds of other particles being characterized by the same matrix element, see for example the random interaction matrix in [92]. As discussed later in more detail, the many-body level density is a fast growing function of excitation energy due to the pure combinatorial reasons and the states with close energy and the same exact quantum numbers are naturally becoming strongly mixed without any artificial randomness. This is a universal source of *quantum chaos*, one of our main future topics.

Although the angular momentum J of the whole system and total isospin T are conserved by the Hamiltonian (or can be deliberately violated to study the consequences), in practice it might be too computationally expensive to formulate the Hamiltonian of the system forming correct coupled rotational superpositions with the help of the $3nj$-symbols. It may turn out better to minimize the computational problems by working in the M-scheme without introducing explicitly the full machinery with high-order coefficients of vector coupling. Even if the total dimension of the matrix increases, this still can be profitable for practical calculations. Of course, the correctness of computation has to be confirmed by the appropriate degree of degeneracy of resulting stationary states.

3.3. Results of the Shell Model

To illustrate the efficiency of the shell model and its reasonable agreement with experimental data we limit ourselves by few examples. Here, in Fig. 3.1

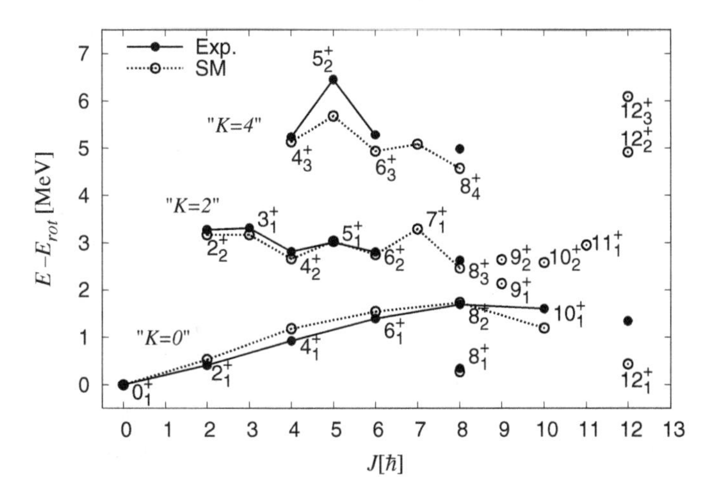

Fig. 3.1. Excitation energies in ^{24}Mg shell model predictions are compared with experiment. The energies are shown relative to the rotor energies defined as $E_{\rm rot} = 0.16J(J+1)\,$MeV. Rotational bands are highlighted and labeled with the K quantum number.

we show a good qualitative and frequently even quantitative agreement of a well-formulated shell-model scheme with experimental data at relatively low excitation energy where in many cases high quality data are available.

Many results of the broad shell-model study for ^{28}Si can be found in the review article [260]. Being exactly in the middle of the *sd* orbital space, this nucleus provides one of the best examples of successful shell-model calculations. In particular, it can be noticed that the energies of the lowest ten 0^+ states are in a good correspondence with the experimental measurements.

3.4. Level Density

The behavior of the level density is indispensable for evaluation of numerous nuclear reactions, including those in nuclear technology and in the astrophysical environment. The main unavoidable feature of the energy spectra in a realistic many-body system is the obvious growth of the level density [267] as a function of the total energy. Here we refer either to the total level density or to individual sectors of the Hilbert space where all levels have the same exact quantum numbers.

In the mean-field basis, the stationary states are formed from certain partitions (distributions of valence nucleons over single-particle levels). For not too small particle numbers, we have, already inside a single partition, many allowed by the Fermi statistics states with the same total spin J, parity Π, and isospin T. The interaction mixes the states within a given partition, as well as between the partitions but, even with no residual interaction, there are many states of the same class around given (not too low) energy. The presence of pairing pushes particle and hole excitations above the pairing gap which makes the level density to strongly increase at energy 1.5–2 MeV. The collective states of vibrational or rotational nature could be shifted by corresponding residual interactions to lower energy also increasing the level density. This situation is a fruitful field for the combinatorics that forces the level density in each class to grow roughly exponentially with excitation energy. The collision-type interactions become then effectively strong acting between the states of the same class located close in energy. Such processes also make the resulting level density more smooth as a function of energy.

Until now we do not have a full analytically derived theory for calculating the many-body level density without straight diagonalzation. Two semi-empirical approaches provide reasonable results in agreement with available experimental information that comes mainly from few practical sources: data on individual levels at relatively low energy, statistics of neutron resonances at excitation energy 6–8 MeV, and photoabsorption experiments which can provide knowledge in a broad interval of energies (the so-called *Oslo method* [116, 117, 236]).

By its nature, the level density $\rho(E)$ is a statistical notion. It is formally defined through the sum of delta-functions around given energy E and for fixed other global parameters $I^{(i)}$ common for the selected class of levels $|\alpha\rangle$,

$$\rho(E, I^{(i)}) = \sum_{\alpha} \delta(E - E_{\alpha}) \prod_{i} \delta(I^{(i)} - I^{(i)}_{\alpha}). \tag{3.2}$$

Of course, only after averaging over some interval of energy, the value of this formally defined highly singular function acquires certain meaning. This makes sense when the observable properties of neighboring states with the same global quantum numbers contributing to the density of a given subclass are quite similar. Otherwise one can have very large fluctuations. This averaging is naturally produced by the mixing residual interactions.

Historically, the Darwin–Fowler method borrowed from statistical mechanics of macroscopic systems was used for the total level density as a function of energy irrespectively of the values of other constants of motion. The detailed derivations can be found in many textbooks on statistical physics including the nuclear textbook [265], see also the review paper [267] with numerous specific references. The main part of the answer is the exponential factor e^S where S is the equilibrium entropy found by the statistical description of the system. This is the point where one needs to specify the model description of the nucleus and it is possible to introduce the nuclear *thermodynamic* temperature through

$$T_{\text{t-d}} = \left[\frac{\partial S}{\partial E}\right]^{-1}. \tag{3.3}$$

Then one can reformulate the approach for special classes of states.

In statistical applications to nuclear structure, the angular momentum dependence of the level density, going back to Bethe [39], is usually introduced through the assumption of the *random coupling* of individual angular momenta of particles as the total nuclear angular momentum $\mathbf{J}$ results in a many-body system from averaging over numerous coupling paths. In the M-scheme, where the total spin projection $J_z = M$ can come from many combinations $M = \sum_a m_a$, this random coupling leads to the Gaussian distribution

$$w(M) = \frac{\rho(E;M)}{\rho(E)} = \sqrt{\frac{\alpha}{\pi}}\, e^{-\alpha M^2} \tag{3.4}$$

with the width $\sigma = 1/\sqrt{2\alpha}$. The level density for a given total nuclear spin J can be found from the logical identity,

$$\rho(E,J) = \rho(E;M=J) - \rho(E;M=J+1). \tag{3.5}$$

There are several ways to predict the level density [85, 267]. The simplest road uses again the Darwin–Fowler method enriched with the roughly described pairing effects. Essentially this is based on the Fermi-gas picture with the so-called *back shift* [101] that in average takes care of the existence of the pairing gap in the spectrum of elementary excitations. The corresponding phenomenological expression is given by

$$\rho(E,M) = \frac{1}{12\sqrt{2}a^{1/4}(E-\Delta)^{5/4}\sigma}\, e^{2\sqrt{a(E-\Delta)}-M^2/2\sigma^2}. \tag{3.6}$$

Here the characteristic parameter a comes originally as a density of single-particle states from the elementary picture of the perfect Fermi-gas of nucleons. In practice this is, along with the back shift energy Δ, a fitted phenomenological parameter.

Currently, one of the broadly used by practitioners phenomenological models for the level density is the so-called *constant temperature model* (CTM) with a pure exponential behavior,

$$\rho(E) = \rho_0 e^{E/T}, \tag{3.7}$$

where the normalization constant is usually written as $\rho_0 = e^{-E_0/T}/T$. The parameters T and E_0 are specific for different nuclei and for various sectors (J,T) in the spectrum of each nucleus. The shell model results are typically in a good agreement with this parameterization providing specific values of parameters for a given nucleus. For all nuclei described by the *sd*-shell model, these parameters are calculated from the microscopic Hamiltonian and tabulated in [132].

We have to stress that the exponential growth (3.7) cannot continue too long as it contradicts to the condition of statistical equilibrium in usual statistical physics. The name of *constant temperature* T is popularly given to this parameter while the thermodynamic temperature is related to this quantity [267] as

$$T_{\text{t-d}} = [1 - e^{-E/T}]T. \tag{3.8}$$

Therefore, the model parameter T plays here the role similar to the *limiting temperature* in high-energy physics that was considered due to the exponential accumulation of resonances as a function of energy. In fact, the form (3.7) loses its validity at some energy when it matches the statistical temperature defined by thermodynamic entropy. We will later see that the tempting interpretation of the CTM [163, 164] as a description of the phase transition related to the melting of superfluid pairs does not work.

3.5. Strength Function

Here we consider a class of states with the same exact quantum numbers. In the presence of a residual interaction, the stationary combinations are superpositions of initially selected basis states. We start with the partitions of the shell model constructed for the independent particle system. This mean-field picture provides the Pauli-allowed fillings of the mean-field states; these many-body states we label $|k\rangle$. With the interaction gradually switched on, those states evolve into more and more complicated superpositions.

The trajectories of evolving energy levels as a function of the interaction strength seem to cross each other. But this in fact does not happen: the crossings are *avoiding*. The evolution of the strength serves as a perturbation repelling, according to standard quantum perturbation theory, the neighboring energy eigenvalucs. As a result of many avoided crossings, we come (even before the full realistic interaction strength is achieved) to a more or less ordered ladder of levels, a *disordered crystal* with relatively small fluctuations of the level spacings, recall Fig. 2.11.

Every final state $|\alpha\rangle$, an eigenstate of the full nuclear Hamiltonian with energy E^{α}, is an intricate combination (2.9) of many unperturbed states $|k\rangle$,

$$|\alpha\rangle = \sum_{k} C_k^{\alpha}|k\rangle, \quad \sum_{k} |C_k^{\alpha}|^2 = 1. \tag{3.9}$$

For the interaction Hamiltonian invariant under time reversal, the amplitudes C_k^{α} can be taken real. The inverse transformation, using the same amplitudes,

$$|k\rangle = \sum_{\alpha} C_k^{\alpha}|\alpha\rangle, \quad \sum_{\alpha} |C_k^{\alpha}|^2 = 1, \tag{3.10}$$

allows one to introduce the *strength functions* of the non-interacting states $|k\rangle$ evolving into realistic states of the fully interacting system,

$$F_k(E) = \sum_{\alpha} |C_k^{\alpha}|^2 \delta(E^{\alpha} - E), \qquad \int dE\, F_k(E) = 1. \tag{3.11}$$

This shows the redistribution of the strength of the original state $|k\rangle$ over the actual energy scale.

An equivalent formal definition of the strength function is

$$F_k(E) = \langle k|\delta(E-H)|k\rangle = -\frac{1}{\pi}\,\mathrm{Im}\left\langle k \left| \frac{1}{E - H + i\eta} \right| k \right\rangle. \tag{3.12}$$

Here we use the known formal identity valid for real x and $\eta \to +0$,

$$\frac{1}{x \pm i\eta} = \mathrm{P.v.}\frac{1}{x} \mp i\pi\delta(x), \tag{3.13}$$

where the symbol P.v. means that all future integrals with this function are to be taken in the sense of the principal value. If the neighboring states in the region of high-level density have similar structure, we can take the average of their weights over the interval of energy around E,

$$F_k(E) \approx \overline{|C_k^{\alpha}|^2} \sum_{\alpha} \delta(E^{\alpha} - E) = \overline{|C_k^{\alpha}|^2}\, \rho(E). \tag{3.14}$$

This defines the practical meaning of the smoothed strength function $F_k(E)$ as the *local density of states* that usually carries a shortened name LDOS in condensed matter theory. The probabilistic meaning of the strength function allows one to define the mean energy of the wave packet corresponding to the initial state $|k\rangle$,

$$\overline{E_k} = \int dE F_k(E) E = \sum_{\alpha} |C_k^{\alpha}|^2 E^{\alpha}, \tag{3.15}$$

and the spread of this energy,

$$\sigma_k^2 = \int dE F_k(E)(E - \overline{E_k})^2. \tag{3.16}$$

It follows immediately that the centroid,

$$\overline{E_k} = \sum_{\alpha} \langle k|\alpha\rangle\langle\alpha|H|\alpha\rangle\langle\alpha|k\rangle = H_{kk}, \tag{3.17}$$

is the diagonal matrix element in the unperturbed basis of the full Hamiltonian. In the same way we can express the width (3.16) of the strength function for a certain unperturbed state,

$$\sigma_k^2 = \sum_{k' \neq k} H_{kk'}^2, \tag{3.18}$$

the sum of squared off-diagonal matrix elements. Therefore, the lowest *moments* of the strength function do not require any diagonalization work and can be directly read from the original Hamiltonian matrix. In fact, all moments can be found in such a way but the amount of work related to the multiplication of huge matrices grows very fast for higher moments.

The actual shape of the strength function evolves with the change of the intensity of interaction mixing the unperturbed states. Here we show the so-called *standard model* [43] that is practically useful in realistic nuclear applications. Let us look at the spreading process of a certain unperturbed state $|k\rangle$ that will be numbered as $|1\rangle$. First we delete the column and the row including this state in the Hamiltonian matrix and assume that the remaining matrix of dimension $N-1$ is diagonalized providing the intermediate states $|\nu\rangle$ with energies ϵ_ν and couplings

$$V_{1\nu} = V_{\nu 1}^* = \sum_{k \neq 1} H_{1k} \langle k|\nu\rangle \tag{3.19}$$

to our special state $|1\rangle$. After this pre-diagonalization, the matrix has the angular element $\overline{E_1} = H_{11}$, interim energies ϵ_ν, $\nu = 2, \ldots, N$, and off-diagonal elements $V_{1\nu}$. Now it is easy to arrive at the final diagonalization that leads from the states $1, \ldots \nu, \ldots$ to eigenstates $|\alpha\rangle$ and provides their energies,

$$E^\alpha = \overline{E_1} + \Phi(E^\alpha), \quad \Phi(E) = \sum_\nu \frac{|V_{1\nu}|^2}{E - \epsilon_\nu}. \tag{3.20}$$

The final squared amplitudes (weights) of the stationary states are

$$|C_1^\alpha|^2 = \frac{1}{1 - [d\Phi/dE]_{E=E^\alpha}}. \tag{3.21}$$

To justify this procedure we assume that the exclusion of a single state does not modify appreciably the statistical properties of the system. The behavior of the function $\Phi(E)$ between the poles is similar to what was shown in Fig. 2.9. If all spacings D between the poles are of the same order (the spacing distribution will be discussed later in more detail), we can model the relevant local level density as $\rho \approx 1/D$. We substitute the squared matrix elements (3.19) by a characteristic constant V^2. Then we come to the average expression for the matrix elements (3.21) as a function of the pole energy,

$$|C(E)|^2 = \left\{1 + \frac{\pi^2|V|^2}{D^2}\left[1 + \cot^2\left(\frac{\pi E}{D}\right)\right]\right\}^{-1}. \tag{3.22}$$

Here the known series was used,

$$\cot\alpha = \sum_{n=-\infty}^{\infty} \frac{1}{\alpha - n\pi}, \quad \text{so} \quad \Phi(E^{\alpha}) \approx \frac{\pi^2|V|^2}{D}\cot\left(\frac{\pi E}{D}\right). \tag{3.23}$$

The convenient notion characterizing the mixing process is the *spreading width*,

$$\Gamma = 2\pi\,\frac{|V|^2}{D}, \tag{3.24}$$

in analogy to the golden rule expressing the quantum probability of a process. Finally, the model leads to

$$|C_1(E)|^2 = \frac{D}{2\pi}\,\frac{\Gamma}{(E - \overline{E_1})^2 + \Gamma^2/4}, \tag{3.25}$$

and the typical strength function of the Breit–Wigner shape

$$F_1(E) = \frac{1}{2\pi}\,\frac{\Gamma}{(E - \overline{E_1})^2 + \Gamma^2/4}. \tag{3.26}$$

As illustrated in Fig. 3.2, despite approximations used in deriving this result, the average realistic strength distribution is still well approximated by the Breit–Wigner shape.

In spite of visible resemblance of Eq. (3.24) to the perturbation theory (square of the mixing matrix element divided by the level spacing), here the situation is quite different. The whole consideration implies the possibility of replacement of the picket fence of levels by a smooth function of energy which means that $V \gg D$, and therefore $\Gamma \gg D$.

With rather brave estimates used in the derivation of Eq. (3.26), the result usually works well in statistical sense. Individual strength functions may have different shapes which are not smooth at all. However, averaging exact shell-model results over few states at close energies quickly provides [94, 260] a typical bell-shape strength function. At not very high excitation energy, this resulting shape is indeed close to the Breit–Wigner formula. But if the resulting width Γ grows, it starts violating the conditions of validity of the approximations made along the road. Gradually one comes to the chaotic limit where the typical shape of the strength function is

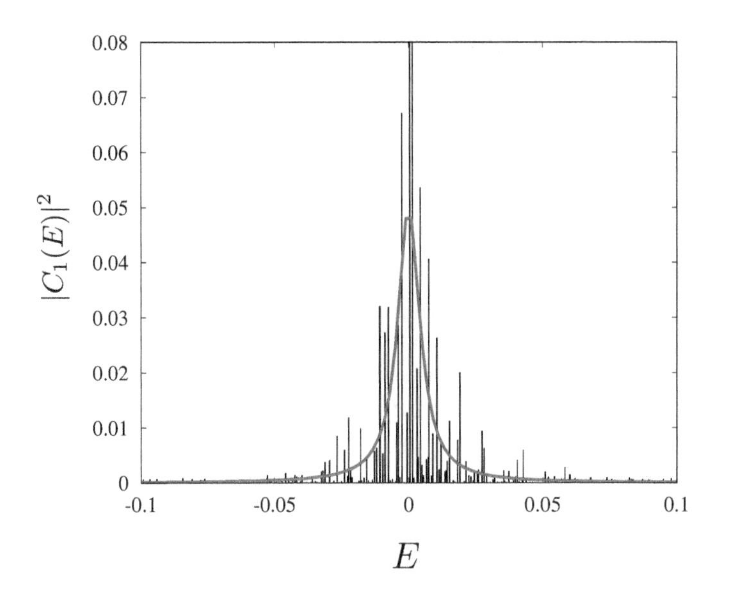

Fig. 3.2. Distribution of strength as a function of energy. Solid red line shows the Breit–Wigner distribution in Eq. (3.25) which is compared with the numerically obtained distribution of coefficients $|C_1(E)|^2$ for a state in the middle of the spectrum where coupling matrix elements in Eq. (3.19) are randomly generated assuming a normal distribution. A randomly generated Gaussian Orthogonal Ensemble which is discussed in chapter 4 is used to generate the spectrum of $N = 4000$ levels distributed in semicircle of radius 2, this defines the energy scale and the average level spacing π/N.

getting close to a Gaussian with faster decreasing wings and finite mean square width that becomes essentially constant in a given excitation region. A certain confirmation of this transformation can be found, apart from numerical estimates, in the phenomenon of double dipole giant resonances (the second harmonic of the universal collective vibration discussed earlier). If the continuum contribution is still not too large, the resulting width Γ_2 of the double resonance is [148] simply a convolution of two single widths. This leads to $\Gamma_2 = 2\Gamma$ in the Breit–Wigner case and $\Gamma_2 = \sqrt{2}\Gamma$ in the Gaussian case. The experiment confirms the second possibility and therefore prefers the Gaussian limit.

3.6. Information Entropy and Effective Temperature

The complexity of individual stationary wave functions is traditionally measured by the information (Shannon) entropy. Here we use only the weights $w_k^\alpha = |C_k^\alpha|^2$ as probabilities of individual "events", namely the non-zero contributions of basis states $|k\rangle$ to the actual event $|\alpha\rangle$. The resulting entropy of a certain stationary state is defined as

$$S^{\alpha} = -\sum_{k} w_k^{\alpha} \ln(w_k^{\alpha}). \tag{3.27}$$

The natural and legitimate objection against using such a quantity as a measure of probabilistic processes in a complicated quantum system is that it obviously depends on the choice of the basis $|k\rangle$. In the actual stationary basis $|\alpha\rangle$, any eigenstate has just a single component, and entropy of each eigenstate measured according to Eq. (3.27) vanishes (no visible complexity in this basis). Agreeing with this statement, one can still deem necessary to quantify the dynamic evolution of the system in the process of switching on interparticle interactions. If the original mean-field basis (or energy density functional) takes proper care of average regular dynamics, the spreading in the interaction processes contains the stochastic components leading to the growth of entropy (3.27) with respect to the original "simple" basis. This evolution is similar to the establishment of thermal equilibrium as will be claimed below.

Reliable versions of the shell model in a finite orbital space, for example [260, 265], allow the exact numerical solution (diagonalization) and show a typical picture of information entropy for individual states changing as a function of excitation energy, Fig. 3.3. The entropy is given by the Gaussian-type strip of a relatively narrow transverse width that becomes even more narrow with the increased dimension of a class of states. One can say that the information entropy of individual states is in fact a rather well-defined function $S(E)$ of energy, and therefore a *thermodynamic function*, with a good particle–hole symmetry with respect to the middle energy in application to a finite shell space.

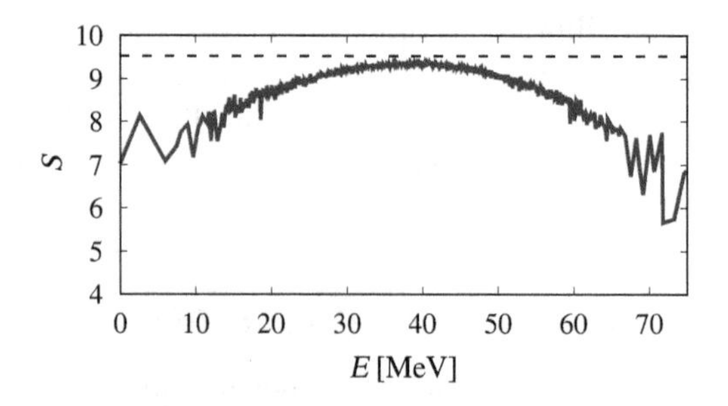

Fig. 3.3. Information entropy for the states with angular momentum $J = 2$ in ^{24}Mg. Dashed line shows the GOE limit of $\ln(0.48N)$, where the dimension $N = 28503$.

Similarly, the level density is converted, after averaging, from a fence of delta peaks into a continuous function $\rho(E)$. A reasonable Gaussian parameterization,

$$S(E) = \frac{1}{\sqrt{2\pi\sigma^2}} e^{-(E-E_c)^2/(2\sigma^2)}, \tag{3.28}$$

allows us to introduce *thermodynamic temperature*,

$$T_{\rm t-d} = \left(\frac{\partial \ln S}{\partial E}\right)^{-1} = \frac{\sigma^2}{E_c - E}, \tag{3.29}$$

that is positive below the centroid $E = E_c$ and negative after that reaching infinity at the centroid. Such a description is not very good at the edges which, regrettably, might be just a region of special theoretic and experimental interest with strong non-statistical effects. There one has to rely on the explicit quantum-mechanical solution.

For a closed mesoscopic system with no external heat bath, effective thermalization comes mainly through the incoherent collision-like interactions. Averaging the results of such processes, we can look at the mean occupation numbers of individual orbitals $|\lambda)$ for each exact stationary state $|\alpha\rangle$. They can serve as possible intrinsic thermometers if the particle occupancies n_λ can be effectively described by the Fermi distribution

$$n_\lambda(T_*) = \frac{1}{1 + \exp[(\epsilon_\lambda - \mu_*)/T_*]}, \tag{3.30}$$

with parameters of single-particle temperature T_* and chemical potential μ_*. It turns out [260] that the effective temperature T_* found from individual stationary states practically coincides as a function of energy with thermodynamic temperature $T_{\rm t-d}$ defined in the preceding paragraph, Fig. 3.4. Similar results were found for complex atoms [92]. A qualitative conclusion is that the interparticle interaction in an isolated system plays the role of the external bath statistically redistributing the particles over single-particle states.

The single-particle thermometer works well as in this example if the interaction strength is coordinated with the parameters of the mean field. For example, in the case of ^{28}Si with the half-filled *sd* orbital space, the centroid of the spectrum, and therefore infinite temperature, exactly coincides with half-filling of all three orbitals. However, if the starting mean field is not consistent with the residual interaction, the thermal picture is strongly distorted [260]. In Figs. 3.5 and 3.6 we show the same study

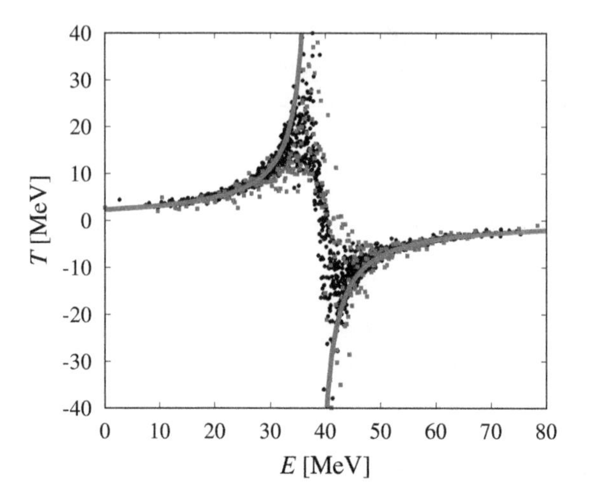

Fig. 3.4. Comparison of single-particle (scattered dots) and thermodynamic (solid line) temperature of individual many-body states in ^{24}Mg; $J = 0$, $T = 0$ states are shown with blue squares and $J = 2$, $T = 0$ states with black circles. Both temperatures are given in a function of excitation energy [265].

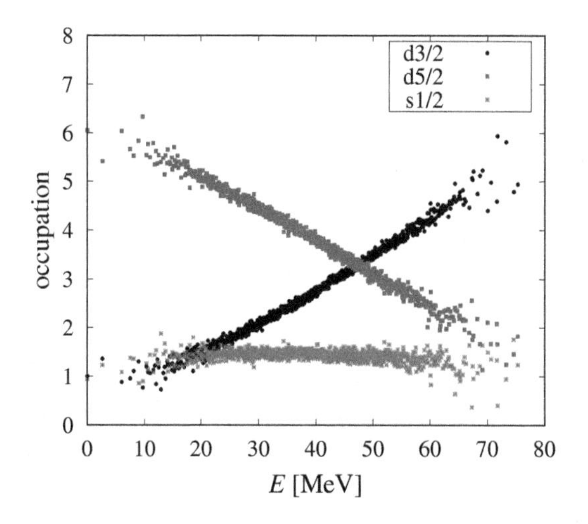

Fig. 3.5. Occupation of orbitals in ^{24}Mg, $J = 2$, $T = 0$ states as a function of energy.

for ^{24}Mg. Figure 3.5 shows occupancies for realistic interaction strength, while Fig. 3.6 corresponds to the case when the strength of all parts of the interaction is artificially made ten times stronger. Then a normal thermodynamic evolution is absent as the mean field is inappropriate.

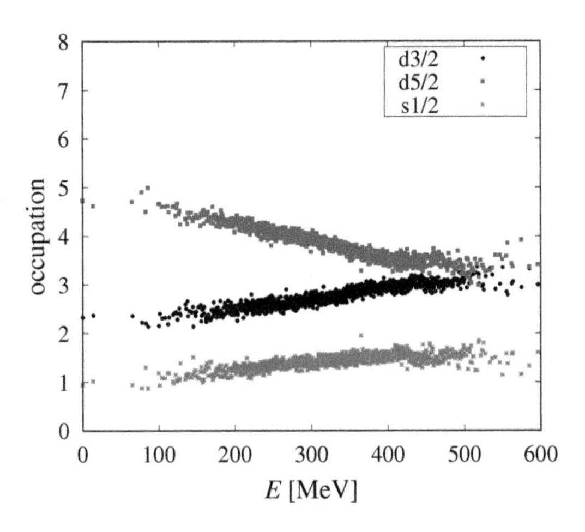

Fig. 3.6. Occupation of orbitals in ^{24}Mg $J = 2$, $T = 0$ states as a function of energy for the two-body interactions enhanced by a factor of ten thus suppressing the mean field.

Another possible measure of the complexity of stationary states is the *inverse participation ratio* also called the number of principal components $N_{\text{p.c.}}$. This is a characteristic of the structure of a state $|\alpha\rangle$ defined as

$$N^{\alpha}_{\text{p.c.}} = \left(\sum_k |C^{\alpha}_k|^4\right)^{-1}. \tag{3.31}$$

For a big dimension N and the state $|\alpha\rangle$ of high complexity, the typical amplitudes C are of the order of $1/\sqrt{N}$, and $N_{\text{p.c.}}$ can be of the order of the whole size N of the orbital space.

There is an important question of differentiating between a typical complicated state (3.27) and a collective state like a giant resonance discussed in Sec. 2.8. Both could be complicated superpositions of many mean-field components. The difference is in the *coherence* of these components. Here a special construction of a *phase correlator*,

$$P^{(\alpha)} = \frac{1}{N}\sum_{kk'} C^{\alpha}_k C^{\alpha *}_{k'}, \tag{3.32}$$

can help [10]. In the case of a large correlated superposition we expect this value to be close to 1, while, for a typical chaotic function, it will be of the order $1/N$. As mentioned earlier, collective states found in the random phase approximation or in a similar approximate way are not exact stationary states. They experience spreading over neighboring complicated superpositions with the same quantum numbers (later we will also account for the coupling to the decay channels). As a result, instead of a certain collective excitation we have a spread strength function concentrating this collective mode. The underlying chaotic states now carry a *scar*, a fraction of the collective strength [131]. If these states are located above the decay thresholds, the way of observing this collective strength will depend on the lifetime of the underlying states and on the reaction leading to their excitation.

Chapter 4

Statistical Ensembles

4.1. Briefly About Classical Chaos

Classical chaos is now a well-established rich avenue of physics covering mechanics, electrodynamics, hydrodynamics and classical statistics [259]. The story goes back to Poincarè and many good books are currently available explaining the underlying physics and mathematics. This topic develops tangentially to our main story so here we limit ourselves by few paragraphs and figures just introducing important terms and mentioning the main ideas, see for example a discussion of multiple applications in [205].

Classical chaos appears essentially in all cases when a system of few (starting even with one in a time-dependent environment) degrees of freedom is described by nonlinear equations of motion. Such systems typically display an extreme sensitivity to initial and boundary conditions. The simplest example is given by a billiard with incompatible boundary conditions on the walls. In a "normal" two-dimensional billiard (just a plane rectangular area with elastic reflection of a particle from fixed boundaries), the energy is strictly conserved but the presence of two dimensions requires another characteristic for every trajectory which is here evident — the particle momentum is conserved as seen from a periodic continuation of the real billiard beyond its actual walls. Every trajectory (except for special cases of motion exactly perpendicular to one of the walls) after infinitely many reflections uniformly covers the whole billiard area. This obviously has nothing to do with chaos as, in the picture with multiple mirror continuations of the trajectory beyond real walls, the whole trajectory is just one straight line, Fig. 4.1(a). Here we have a system with the two-dimensional phase space and two constants of motion (energy and wave vector in the sense of its continuation similar to the case of a realistic perfect crystal). A similar situation would take place in a circular billiard where we again

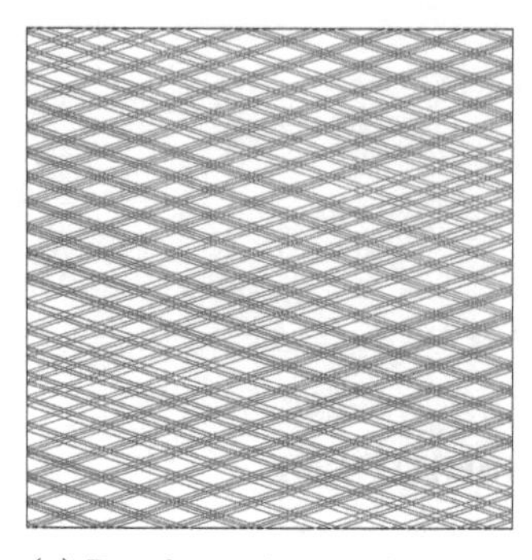

(a) Regular trajectory in an ideal rectangular billided.

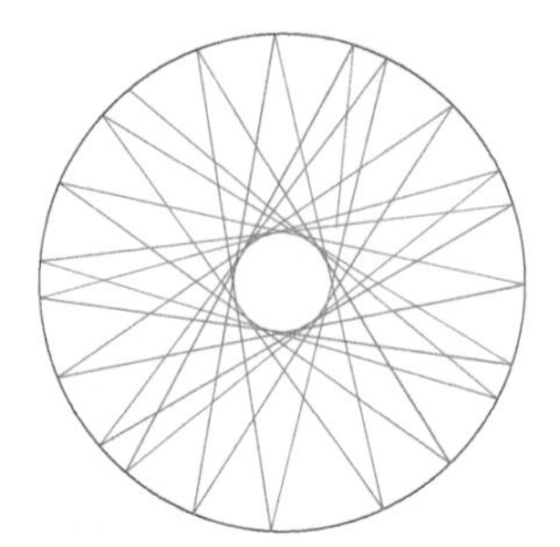

(b) Regular trajectory in an ideal circular billided.

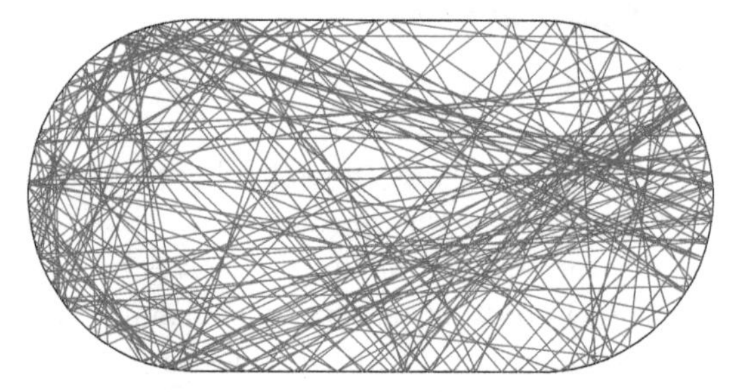

(c) Chaotic trajectory for a stadium billiard.

Fig. 4.1. Rectangular, circular, and chaotic stadium billiards.

have two conserved constants of motion, energy and orbital momentum, see Fig. 4.1(b)

The *stadium billiard*, Fig. 4.1(c), has contradictory boundary conditions, a hybrid of straight and circular walls. Therefore, neither linear nor angular momentum is conserved and energy remains the only available constant of motion. A typical individual trajectory (again except for singular trajectories which form a set of measure zero) covers the *phase space*. A small difference in initial conditions leads, after some time, to completely divergent trajectories. This makes the problem of predicting far enough future motion based on initial conditions practically unsolvable: the trajectories are *unstable* with respect to tiny initial differences, in particular in calculations with finite computer precision after unavoidable round-off errors in the last digits which become visible with time progressing along the trajectory. The exact predictions, strictly speaking, become impossible for sufficiently long-time intervals. The formal solutions of classical equations of motion lose their predictive power, in spite of the fact that seemingly the unique trajectory satisfies (for formally exactly fixed initial conditions!) the equations of motion being given by the least action principle.

A classical model of a periodically kicked rotator is a good example illustrating the emergence of chaotic dynamics in classical mechanics. Consider a particle kicked periodically by a homogeneous force in a fixed direction. As a result of each such impulse, the momentum component p along the force direction is increased by K. In simplified units, the dynamics can be presented in a discretized form as

$$\begin{aligned} p_{n+1} &= p_n - K \sin \theta_n, \\ \theta_{n+1} &= \theta_n + p_{n+1}. \end{aligned} \tag{4.1}$$

The phase portrait of the system can be generated with initial coordinate and momentum values (θ_0, p_0) by iterating Eq. (4.1). The physical location of the particle is defined when θ is taken modulo 2π; it is clear that p can also be taken modulo 2π. The phase portrait showing points (θ_n, p_n) under the periodicity assumption is known as the Chirikov standard map [64, 65]. Figure 4.2 illustrates the phase portrait for a trajectory starting from the initial condition $(\theta_0, p_0) = (1, 0)$. The trajectory (a) corresponds to $K = 2$, while the one on the right (b) is for $K = 2.7$, the difference is significant. For a smaller K, the trajectory remains regular and all points (θ_n, p_n) remain on a single curve, while, for the larger values of the parameter K, the points cover the map uniformly.

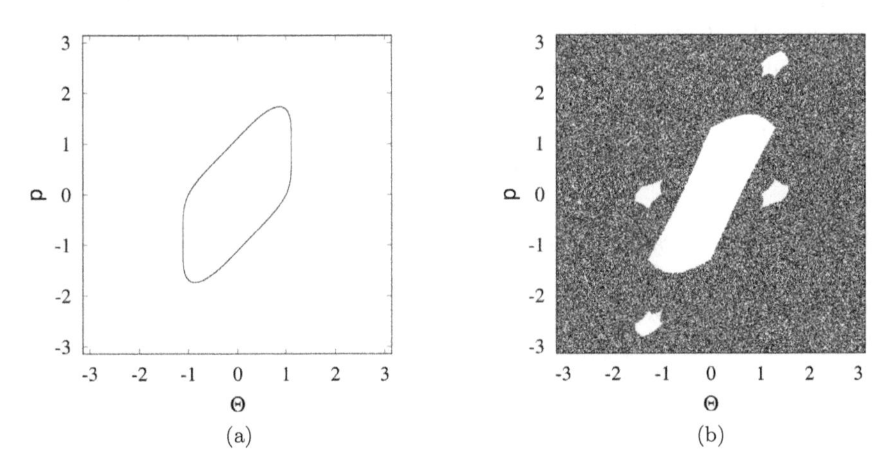

Fig. 4.2. Kicked pendulum, map of a single trajectory with initial condition $(\theta_0, p_0) = (1, 0)$. (a) Trajectory with $K = 2$. (b) Trajectory with $K = 2.7$.

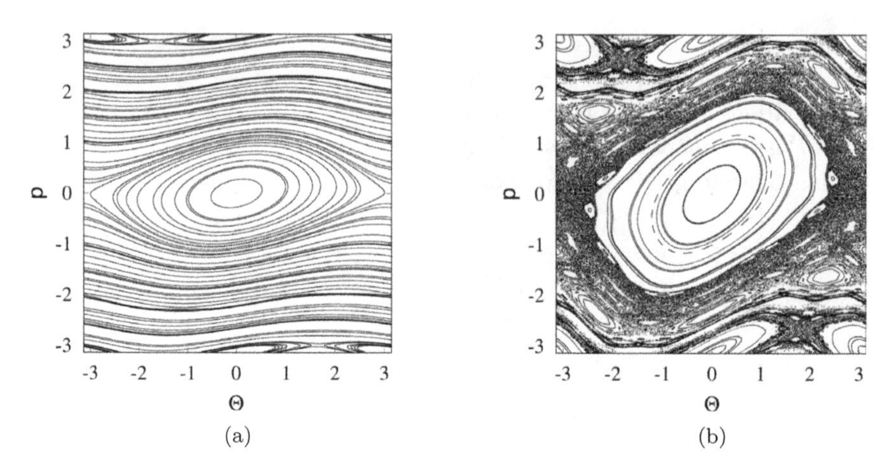

Fig. 4.3. Kicked pendulum, map of multiple trajectories. (a) Trajectories with $K = 0.3$. (b) Trajectories with $K = 1$.

The kicked rotator model with its continuous parameter K allows us to explore the transition to chaotic motion. In Fig. 4.3, we show two maps that represent many trajectories (generated with different initial conditions) being superimposed. This collective map looks very uniform for regular dynamics, panel (a) for $K = 0.3$, while for a larger value, $K = 1$, shown in (b), some trajectories become chaotic while others still remain regular as seen from different areas on the map.

There exist special mathematical families of classical systems which have full sets of analytical constants of motion with their number equal to the total number of degrees of freedom. The Hamilton functions of such systems typically reveal a high degree of symmetry. As a result, the choice of generalized coordinates is possible that leads to separation of variables. Under this choice, the motion along different degrees of freedom is independent; the energy is given by the sum of partial energies corresponding to independent separated variables. Usually this happens when the realistic dynamics of the system are strongly simplified neglecting many realistic factors. Already for few degrees of freedom, the typical physical systems have destroyed constants of motion, except for global energy, total momentum of a system as a whole and total angular momentum. Then phase space trajectories become unstable with respect to small variations of initial conditions. The characteristic motion acquires features of a random process when a phase space trajectory completely fills in certain regions of the hypersurface of constant global quantities.

Traditionally, in classical mechanics and in a more general theory of dynamical systems, an exponential growth of the specifically measured distance between two trajectories very close at the initial time moment is considered as an estimate of the character of motion [152]. This is discussed in many textbooks and in review papers [79, 182]. The number of different exponents is equal to the number of degrees of freedom, and the actual behavior of the system depends on the initial direction in the dynamical space. The largest positive exponent is called the *Lyuapunov exponent* proper. This serves as a measure of the stability of motion and reliability of predictions made with the use of a device of certain precision [235].

4.2. Quantum Chaos

The widespread explanation of quantum chaos as a manifestation of remnants of classical chaos in quantum conditions seems to be made "upside down". The fundamental dynamics proceed on the quantum level and from there the limiting transition to classical physics, usually through the semiclassical regime, can be traced down in certain cases. The nuclear dynamics based on isospin-invariant interactions of fermions in many aspects have no direct classical analogs. The natural line of formulating a fundamental physical theory goes from deeper quantum to upper classical. In some sense, it would be more logical to speak of classical chaotic phenomena as remnants of underlying quantum interactions, if indeed the clear classical limit does exist.

As mentioned above, one can work with special examples of classical systems that can be called *integrable*. Such a system has a complete set of analytic constants of motion the number of which coincides with the number of degrees of freedom. The Hamilton functions of such systems usually have a high degree of symmetry. As a result, it might be possible to introduce generalized coordinates leading to separation of variables. The motions along different degrees of freedom are then independent so that total energy is a sum of corresponding energies. However, such examples are exceptional. Already at a small number of degrees of freedom, the typical situation corresponds to destroyed individual constants of motion, apart from global and in a sense trivial quantities as total energy, linear and angular momentum of motion as a whole. Typically, the phase trajectories acquire features of a random process so that the path integrals have to include interference of many trajectories giving comparable contributions, in contrast to the quasiclassical situation where the least action trajectory provides an overwhelming contribution. With the growth of total energy, this picture becomes standard: the phase trajectories gradually fill in the entire hypersurface of total energy in the phase space.

This process of stochastization is unavoidable in realistic quantum systems. The complexity of a wave function, especially in a many-body quantum system, is born naturally with the growth of energy. Any weak interaction between degrees of freedom breaks their independence and makes the resulting wave function a complicated superposition of initially independent components. It is impossible to derive a complete set of quantum numbers characterizing a given wave function but it might be not needed. Here it is natural to introduce a statistical description based on general regularities standard for such systems. This is the point where the difference between macroscopic and mesoscopic cases can show up. In macroscopic physics, the individual wave functions of a realistic system become useless at some degree of excitation. Only certain characteristics typical for a huge amount of close quantum states can be studied and reproduced experimentally. In the mesoscopic regime, the individual states still can be studied with high degree of precision.

As a historically important example in nuclear physics, we can mention the neutron resonances in medium and heavy nuclei. In a mesoscopic system as a nucleus, each resonance usually can be resolved and studied in the low-energy region where the resonances are long-lived and non-overlapping. In a macroscopic piece of the substance made of the same chemical elements and therefore nuclei, we can only use the language of the neutron refraction index and related observables. Here the details of probably chaotic neutron trajectories in the medium become hidden being substituted by statistical characteristics.

The advantage and difficulty of mesoscopic physics are in a unique combination of individual wave functions and statistical elements coming from quantum chaos that exists already on the level of specific quantum states. Even if individual quantum levels could be in principle resolved and studied (for neutron resonances this might be necessary), we have to understand general regularities which appear under such conditions. Then it is appropriate to use the statistical language different from and complementary to standard statistical physics of large systems. In the limit of generalization, there exist certain regularities that are generated by fundamental physical principles, such as preservation of probability (unitarity), Hermiticity of the Hamiltonian, time-reversal invariance or its violation, etc. Such general regularities are manifested in statistical properties of families of stationary quantum states at close energy and with identical exactly conserved global quantum numbers. Here we speak about mesoscopic systems at high-level density when even relatively weak interactions are getting effectively strong and make as an output of multiple perturbations an extremely complicated wave functions, while the ensemble of such functions acquires statistical regularities.

The necessary mathematic language for description of quantum chaos can be based, up to some extent, on theory of random matrices [51, 155, 186]. The whole theory was started in classical works by Wigner [250] and Dyson [78]. This marks a transition to even greater degree of abstraction than in the thermodynamical Gibbs ensemble, although we will see a certain connection between these two statistical approaches (recall, for example, paragraphs on entropy and temperature in the previous section). In a sense, in this limiting theory we not only refuse to give a detailed description of the system but we are looking for regularities existing independently of the nature of the system that belongs to a wide class of similar systems.

We can assume a *black box* full of particles interacting in some (unspecified) way that leads to the limiting complexity of many-body eigenstates. This is equivalent to averaging not over the Gibbs ensemble but rather over the ensemble of Hermitian Hamiltonians within the same class with respect to the most general symmetries. The main assumption is that of *ergodicity* that makes the ensemble averaging equivalent to the averaging over a typical fixed Hamiltonian matrix. We start with Hermitian matrices of a definite broad class of stable many-body quantum systems and then compare the ensemble predictions with the properties of realistic individual examples. Later we come to unstable systems and non-Hermitian Hamiltonians.

4.3. Gaussian Random Matrix Ensembles

A random matrix ensemble [43, 155, 186] consists of matrices of a certain dimension with matrix elements of the Hamiltonian operator $\hat{H}$ generated according to some probabilistic recipe. An ensemble is specified by the distribution function $P(\hat{H})$ of matrix elements with a measure $d\mu(\hat{H})$ in the space of those matrix elements. Gaussian ensembles introduced by Wigner [250] occupy a special place in the world of random matrices. Here, the general Gaussian probability is defined globally by the trace of the Hamiltonian,

$$dP_\beta(\hat{H}) = e^{-\beta(N/a^2)\mathrm{Tr}(\hat{H}^2)}\, d\mu_\beta(\hat{H}). \tag{4.2}$$

The parameter β distinguishes possible specific types of the ensembles defined by the characteristic symmetries of matrix elements; N is a dimension that can be increased to infinity, and a is the only parameter that characterizes the actual width of the eigenvalue spectrum. The center of the spectrum for simplicity is arbitrarily placed at the origin of the energy scale.

In nuclear applications, one usually considers the Hilbert space of a large dimension $N \gg 1$ with a Hermitian Hamiltonian $\hat{H}$. In each realization, the stationary states $|\alpha\rangle$ can be expressed as superpositions (2.5) of basis states $|k\rangle$. The Hamiltonian matrix elements depend on the basis. The main and very strong assumption is that the statistics of matrix elements $H_{kk'}$ are invariant under the change of the basis by unitary transformations. Under this assumption, the statistics should be defined in terms of invariants like the trace of the Hamiltonian matrix. Of course, in the eigenbasis $|\alpha\rangle$ of $\hat{H}$, the matrix is pure diagonal with all off-diagonal elements absent. But such bases form a set of null measure in the Hilbert space.

Below we discuss the main properties of Gaussian ensembles which are presented by the matrices of a typically large dimension with randomly taken matrix elements where only some very general symmetry properties are imposed. For a realistic system of N particles, the Hamiltonian matrix is not uniformly full. In nuclear and solid-state applications, the main role is played by the pair-wise ("two-body") interactions. Physically it is partly justified by the fact that a significant fraction of the many-body interactions is taken into account by the mean field ("one-body" part of the resulting Hamiltonian) defined as a background for all secondary interaction processes. Therefore, it is possible to introduce the so-called *embedded random matrix ensembles*, see the review [138]. Here the interaction part of the Hamiltonian (on top of the nuclear mean field) is presented by random matrices, for example of a two-body type but respecting the conservation laws of the corresponding shell model. Many predictions of this approach are very close to those of the full Gaussian ensembles but now they can be directly compared with the specific theory of the system (like nuclear shell-model results). The genuine collective motion can also appear here with some probability.

The assumption that the Hamiltonian is Hermitian means that matrix elements satisfy $H_{mn} = H^*_{nm}$. Making an additional assumption that the system is invariant under time reversal, we can make all matrix elements real and therefore the matrix symmetric, $H_{mn} = H_{nm}$. Such a matrix is defined by $(1/2)N(N+1)$ random statistically independent real parameters.

In this case, we set $\beta = 1$ and the measure in this space is defined by the volume element

$$d\mu_1(\hat{H}) = \Pi_{m \leq n} dH_{mn}. \tag{4.3}$$

Under such a definition, the resulting *Gaussian Orthogonal Ensemble* (GOE) corresponds to the most broad freedom of choice maximizing [260] the ensemble *entropy* defined as

$$S_{\text{ens}} = -\int dH\, P(H) \ln P(H) \tag{4.4}$$

under conditions of normalization,

$$\int dH\, P(H) = 1, \tag{4.5}$$

and fixed dispersion,

$$\int dH\, P(H) \text{Tr}[(H - E_0)^2] = \text{const}, \tag{4.6}$$

where the arbitrary centroid E_0 is taken as zero in (4.2), while the parameter N/a^2 fixes the width of the distribution.

Removing the requirement of time-reversal invariance but keeping the condition of Hermiticity, we allow complex matrix elements with the metric in the space of matrix elements as ($\beta = 2$)

$$d\mu_2(\hat{H}) = \Pi_{m \leq n} d[\text{Re}(H_{mn})]\, \Pi_{m<n} d[\text{Im}(H_{mn})]. \tag{4.7}$$

This is the *Gaussian Unitary Ensemble* (GUE) with a distribution function invariant under arbitrary unitary transformations. In applications, the unitary ensemble will be recalled for the search of violation of the time-reversal invariance and in relation to unstable (open) systems which by construction are not time-reversal invariant. In Eq. (4.7) we take into account that the matrices of the GUE have $N(N+1)/2$ parameters for the symmetric real part and $N(N-1)/2$ off-diagonal parameters for the imaginary part.

Finally, we mention (but practically will not use) the *Gaussian Symplectic Ensemble* (GSE), where one keeps time-reversal symmetry $\mathcal{T}$ but assumes $\mathcal{T}^2 = -1$, and the total angular momentum of the system takes half-integer values. The realistic examples can be found for odd-A deformed nuclei in the states without axial symmetry when the intrinsic angular momentum projection K is not a good quantum number. In any allowed basis, such matrices contain degenerate doublets of time-inverted states.

This structure is invariant under transformations of the *symplectic group* [193]. This group [193] that is known to describe [11] the analytic structure of classical Hamiltonian dynamics with pairs of conjugate variables (coordinate and momentum for each degree of freedom); it is also useful [144] in applications to nuclear collective motion. In a random matrix member of this ensemble, one has, for each of N doublets, two pairs of complex conjugate matrix elements, or four real parameters, altogether $N + 4N(N-1)/2 = N(2N-1)$ parameters. The corresponding ensemble index, recall Eq. (4.2), is $\beta = 4$.

The conclusion is that, for such cases of maximum allowed freedom, the global symmetry of the random Hamiltonian determines corresponding matrix classes of universality. Different, but related, statistical ensembles can be constructed, for example, for the scattering matrix [78], where, due to the unitarity of this matrix, the eigenvalues are complex numbers on the unit circle. Taking into account also the continuum irreversible decays, we will introduce additional ensembles.

4.4. Level Statistics

The statistical properties of the energy spectra are the most easily extracted from the experiments if the full spectrum is reliably known; many studies have been done comparing the data with shell-model calculated spectra and with artificial random matrices (an obvious practical problem is the completeness of the experimental spectrum).

The nearest-level spacing distribution inside a class of states with the same global symmetry, for example, the same total nuclear spin, parity and isospin, can be predicted from general arguments for small spacings s. This comes from a simple consideration of interaction between two close eigenfunctions as a function of mixing parameters. Consider two neighboring levels within the same general class. The effective Hermitian Hamiltonian submatrix of this 2×2 block contains two "unperturbed" energies, ϵ_1 and ϵ_2 at the original distance $\Delta = \epsilon_2 - \epsilon_1$, and the mixing matrix element $V_{12} = V_{21}^* \equiv V$. The diagonalization in this subspace leads to the two eigenvalues

$$E_{\pm} = \frac{1}{2}\left[\epsilon_1 + \epsilon_2 \pm \sqrt{\Delta^2 + 4|V|^2}\right] = \frac{1}{2}(\epsilon_1 + \epsilon_2 \pm s), \tag{4.8}$$

where the resulting spacing is

$$s = E_+ - E_- = \sqrt{\Delta^2 + 4|V|^2} \geq \Delta. \tag{4.9}$$

As was established by Wigner and von Neumann in the early years of quantum mechanics (and by Rayleigh even before quantum era), the presence of the mixing interaction V always *repels* the eigenvalues even if they were originally degenerate. In the statistical ensemble with real V, we have $x = \Delta$ and $y = 2V$ as random coordinates in the plane with given distribution functions $X(x)$ and $Y(y)$. The spacing distribution is then

$$P(s) = \int dx\, dy X(x)Y(y)\delta(s - \sqrt{x^2 + y^2}), \tag{4.10}$$

or, introducing polar coordinates (φ, r) in the (x, y)-plane,

$$P(s) = \int d\varphi\, dr\, r\, X(r\cos\varphi)Y(r\sin\varphi)\delta(s-r) = s\int d\varphi\, X(s\cos\varphi)Y(s\sin\varphi). \tag{4.11}$$

If the distribution functions $X(x)$ and $Y(y)$ of these variables are not singular in the limits of zero arguments, the probability of small spacings s is linearly proportional to s,

$$P(s \to 0) = 2\pi X(0)Y(0)\, s. \tag{4.12}$$

In this way, the GOE predicts the *linear* level repulsion at short distances according to the vanishing probability for two random points in a plane to coalesce at the origin. In general, we expect the large sequence of levels with the same exact quantum numbers to remind a disordered crystal with small fluctuations of their spacings around the mean value determined by the average level density. In the GUE with complex perturbations V, the level crossing requires three random quantities — the original distance Δ, real and imaginary parts of the coupling V — to coincide at the origin. The probability of this goes as s^2 at small s being much smaller than for the GOE.

As physically the necessity to switch to the GUE from the GOE comes from the violation of time-reversal invariance, the analysis of statistics of actual nuclear spectra at small spacings, under the assumption of its random character, can give information on the magnitude of time-reversal violating forces in nuclear dynamics. This is amazing from the general viewpoint: the global property of fundamental value can in principle be checked by the local spectral statistics. Unfortunately, just because of the small probability of appearance of close levels with the same global symmetry, such a statistical search for the time-reversal invariance violations in nuclei did not give better results than those known from direct studies of nuclear processes.

In order to reliably evaluate the whole spacing distribution, it is sufficient [43] to consider the smallest matrices, 2×2. Due to the GOE symmetry, this case has only three independent real elements, and the probability of a certain Hamiltonian is the product,

$$P(H) = P_{11}(H_{11})P_{12}(H_{12})P_{22}(H_{22}). \tag{4.13}$$

This probability should be invariant under orthogonal transformations, here just under rotations in the two-dimensional plane,

$$|1'\rangle = \cos\theta|1\rangle - \sin\theta|2\rangle, \quad |2'\rangle = \sin\theta|1\rangle + \cos\theta|2\rangle. \tag{4.14}$$

The matrix elements of the Hamiltonian in the new basis are bilinear functions of $\sin\theta$ and $\cos\theta$. Now we require the invariance of the distribution function (4.13). With an infinitesimal rotation, this condition can be derived equating the linear in θ terms in the distribution function (4.13) and separating variables H_{ij},

$$(H_{11} - H_{22})\frac{1}{P_{12}}\frac{dP_{12}}{dH_{12}} + 2H_{12}\left(\frac{1}{P_{22}}\frac{dP_{22}}{dH_{22}} - \frac{1}{P_{11}}\frac{dP_{11}}{dH_{11}}\right) = 0. \tag{4.15}$$

The solution of this set of equations by separation of variables is given by Gaussian functions with two separation constants. After the normalization, the result is

$$P_{11} = \sqrt{\frac{C}{4\pi}}\, e^{-C(H_{11}-E_0)^2/4}, \quad P_{22} = \sqrt{\frac{C}{4\pi}}\, e^{-C(H_{22}-E_0)^2/4} \tag{4.16}$$

for the diagonal matrix elements, and

$$P_{12} = P_{21} = \sqrt{\frac{C}{2\pi}}\, e^{-C(H_{12})^2/2} \tag{4.17}$$

for the off-diagonal elements. Confirming the invariance of the result, one can express the total probability (4.13) in terms of the trace,

$$P(H) = \text{const}\, e^{-(C/4)\text{Tr}[(\hat{H}-E_0)^2]}. \tag{4.18}$$

The choice of the centroid E_0 is irrelevant, one can set $E_0 = 0$. The direct generalization for matrices of an arbitrary dimension N leads, after normalization and counting diagonal and off-diagonal matrix elements, to

$$P(H) = \left(\sqrt{\frac{C}{4\pi}}\right)^N \left(\sqrt{\frac{C}{2\pi}}\right)^{N(N-1)/2} e^{-(C/4)\text{Tr}(H^2)}. \tag{4.19}$$

For all following considerations, it is sufficient to use the mean values of quadratic combinations of matrix elements,

$$\overline{H^2_{\text{diag}}} = \frac{2}{C}, \quad \overline{H^2_{\text{off-diag}}} = \frac{1}{C}, \tag{4.20}$$

or, in a more formal way,

$$\overline{H_{ij}H_{kl}} = \frac{1}{C}(\delta_{il}\delta_{jk} + \delta_{ik}\delta_{jl}). \tag{4.21}$$

As always for Gaussian distributions, it is necessary to define only the second moment. Everything is expressed through the one-dimensional constant C related to the inverse width of the distribution. The convenient choice is

$$C = \frac{4N}{a^2}. \tag{4.22}$$

With this normalization, the trace

$$\overline{(\text{Tr}\, H)^2} = \sum_{ij} \overline{H_{ii}H_{jj}} = \sum_{ij} 2\delta_{ij}\, \frac{a^2}{4N} = \frac{a^2}{2} \tag{4.23}$$

does not depend on N that can be taken in the infinite limit. The quantity a determines the spread of energy eigenvalues. The two-dimensional system with different a-priori distribution of diagonal and off-diagonal matrix elements can be also solved in a similar way [61].

The result (4.17) can be used for deriving the distribution of spacings between the closest levels. The transformation from the arbitrary original basis with matrix elements H_{11}, H_{22}, H_{12} to the eigenbasis with eigenvalues E_1 and E_2 and rotation angle θ between these two bases is

$$H_{11} = E_1 \cos^2\theta + E_2 \sin^2\theta, \quad H_{22} = E_1 \sin^2\theta + E_2 \cos^2\theta,$$
$$H_{12} = H_{21} = (E_1 - E_2)\sin\theta\, \cos\theta. \tag{4.24}$$

The Jacobian J of this transformation does not depend on the rotation angle,

$$J = \frac{\partial(H_{11}, H_{22}, H_{12})}{\partial(E_1, E_2, \theta)} = E_1 - E_2. \tag{4.25}$$

As expected, this Jacobian vanishes when the levels intersect as at this point the inverse transformation is not defined because one can take any superposition of degenerate states. The resulting probability distribution for 2×2 matrices does not depend on θ, as should be due to the orthogonal invariance,

$$P(E_1, E_2, \theta) = \text{const}\, |E_1 - E_2|\, e^{-(2/a^2)(E_1^2+E_2^2)}. \tag{4.26}$$

After normalizing this distribution function and integrating over angles and absolute position of the levels, we find, as a generalization of (4.12),

$$P(s) = \int d\theta\, dE_+\, dE_-\, P(E_+, E_-, \theta)\delta(s - E_+ + E_-) = \frac{2s}{a^2} e^{-s^2/a^2}, \tag{4.27}$$

It might be convenient to express the level spacing in terms of the mean spacing,

$$D = \int_0^\infty ds\, P(s)s = \frac{\sqrt{\pi}}{2}\, a. \tag{4.28}$$

In realistic applications, the actual level density sharply grows with energy. Therefore, it is convenient to rewrite Eq. (4.27) in terms of the local distance $x = s/D$ that would allow to compare statistical features at different parts of the energy scale,

$$P(x) = \frac{\pi x}{2}\, e^{-(\pi/4)x^2}. \tag{4.29}$$

This distribution is shown in Fig. 4.4. Although it was derived only for the simplest case of dimension $N = 2$, it is quite close to the exact result for an arbitrary dimension [137, 155] and therefore frequently used for comparison to realistic energy spectra. In the same way, in the case of the violated time-reversal invariance, the corresponding result shows quadratic repulsion,

$$P(x) = \frac{32}{\pi^2}\, x^2\, e^{-(4/\pi)x^2}. \tag{4.30}$$

It is necessary to have in mind that this characteristic distribution makes sense only for the collection of levels within the certain class of exact quantum numbers: the mixture of different classes, when the neighboring states may not be able to interact and can simply cross in the process of evolution, leads to the *Poisson distribution* of level spacings [115, 263]. In the limit of many superimposed orthogonal families of levels, we come to

$$P(x) = e^{-x}. \tag{4.31}$$

In both ensembles, GOE ($\beta = 1$) and GUE ($\beta = 2$), the joint distribution of all eigenvalues can be presented in the form

$$P_\beta(E_1, ..., E_N) = \prod_{m<n} |E_m - E_n|^\beta \, e^{-\beta(N/a^2)\sum_n E_n^2}, \tag{4.32}$$

in agreement with the above mentioned rules of the closest revel repulsion. The parameter β is related to the degree of the level repulsion rule at small distances.

It was mentioned that the found universal distributions do not account for the physical growth of the level density with increasing energy. These distributions in fact deal with the fluctuations of spectra rather than with the actual level density that will be discussed later. Therefore, the procedure of *unfolding*, reduction of the given level sequence to that with the universal mean density, is required prior to studying the distribution of the level spacings. It was suggested that the need for unfolding of the energy spectrum can be eliminated with the use of another statistical measure [174], the distribution of *ratios* of consecutive level spacings.

According to [13, 14], this distribution can be defined as follows. Given an ordered, hopefully complete, set of the energy levels, E_n, the nearest-neighbor spacings are derived, $s_n = E_{n+1} - E_n$. The ratio of consecutive

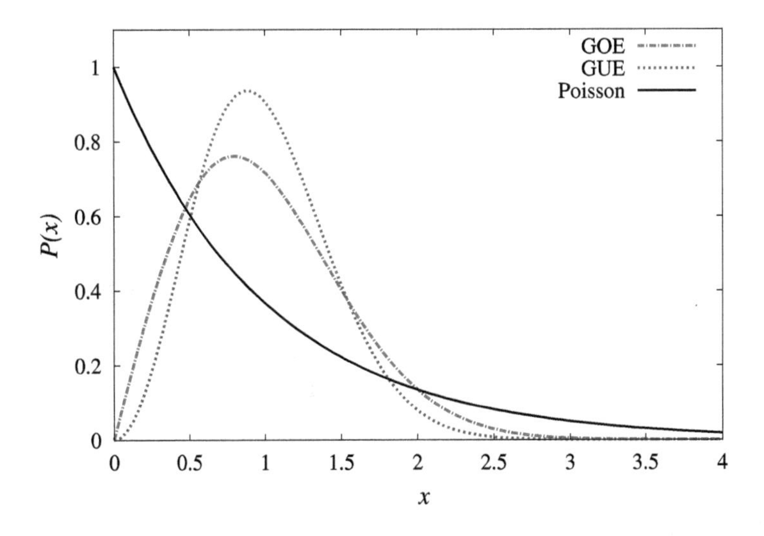

Fig. 4.4. Level spacing distributions for GOE, GUE and Poisson cases.

level spacings is

$$r_n = \frac{s_n}{s_{n-1}}. \tag{4.33}$$

The Poisson limit of the ratio distribution for physically independent consecutive levels was found long ago [115],

$$W(r) = \frac{1}{(1+r)^2}. \tag{4.34}$$

In the same articles [13, 14], the authors derive the ratio distributions for all three canonical random matrix ensembles, orthogonal (GOE, $\beta = 1$), unitary (GUE, $\beta = 2$), and symplectic (GSE, $\beta = 4$),

$$W_\beta(r) = \frac{1}{Z_\beta} \frac{(r + r^2)^\beta}{(1 + r + r^2)^{1+(3/2)\beta}}. \tag{4.35}$$

For the GOE distribution, $\beta = 1$ and $Z_1 = 8/27$ [13]. The transition from the Poisson distribution to that of the GOE was studied [62] for the random matrix (one-body + two-body) ensembles in analogy to similar studies for the level spacing distributions. An application [133] to realistic nuclear many-body models shows that the results using the ratio statistics agree with those derived earlier with the spacing statistics after unfolding.

A simple approximate interpolation between the Poisson and Gaussian limits [51],

$$P_\omega(s) = \alpha(1 + \omega)s^\omega \exp(-\alpha s^{1+\omega}), \tag{4.36}$$

where

$$\alpha = \Gamma[(2 + \omega)/(1 + \omega)]^{1+\omega}, \tag{4.37}$$

can be useful for comparing the intermediate situations on the road to chaos; the value of ω can be found by performing a least square fitting of Eq. (4.36) to the empirical or computational data. This Brody distribution interpolates between the Poisson statistics for $\omega = 0$ and that of the Wigner distribution when $\omega = 1$. The ratio distribution can be written as

$$W(r) = \int ds_n\, P(s_n) \int ds_{n-1}\, P(s_{n-1})\, \delta((s_n/s_{n-1}) - r). \tag{4.38}$$

This is complicated as in principle we have to impose the condition that there are no levels between n and $n-1$. If we forget about that, we can say that, in the sufficiently dense spectrum, $P(s_n)$ and $P(s_{n-1})$ are approximately given by the same function $P(s)$, so that

$$W(r) = \int ds'\, P(s') \int ds\, P(s)\, \delta((s'/s) - r). \tag{4.39}$$

Here $\delta((s'/s) - r) = s\delta(s' - rs)$, then

$$W(r) = \int ds\, s\, P(rs)P(s). \tag{4.40}$$

For the Brody function of Eq. (4.36),

$$W(r) = \alpha^2(1+\omega)^2 r^\omega \int_0^\infty ds\, s^{1+2\omega} \exp[-\alpha s^{1+\omega}(1 + r^{1+\omega})]. \tag{4.41}$$

Here, the integral can be taken exactly with the result

$$W_\omega(r) = (1+\omega)\, \frac{r^\omega}{(1 + r^{1+\omega})^2}. \tag{4.42}$$

In particular, for a limit of $\omega = 0$, we come to the Poisson result (4.34),

$$W_0(r) = \frac{1}{(1+r)^2}, \tag{4.43}$$

while for the GOE limit, $\omega = 1$,

$$W_1(r) = \frac{2r}{(1+r^2)^2}, \tag{4.44}$$

with the qualitatively correct behavior and the maximum at $r = 1/\sqrt{3} = 0.577$.

4.5. Electrostatic Analogy and Relation to Random Processes

Dyson [78] indicated and effectively used an analogy (physical repulsion) between the spectra of random matrices and the properties of the one-dimensional Coulomb gas. Introduce the image of temperature $T = 1/\beta$

and write down the distribution function as

$$P_\beta(E_1,\dots,E_N) \propto e^{-\Phi/T}, \tag{4.45}$$

where the effective potential is

$$\Phi(E_1,\dots,E_N) = -\sum_{m<n} \ln|E_m - E_n| + \frac{N}{a^2}\sum_n E_n^2. \tag{4.46}$$

This function can be interpreted as potential energy of N unit charges located along the straight line $-\infty < E < \infty$; the charges can move only along this line. The first item in the right-hand side of (4.46) corresponds to the Coulomb repulsion of these charges, while the second item gives the potential energy of the charges in the common oscillator field with the equilibrium point at $E = 0$. The level repulsion acquires here the obvious meaning of the Coulomb repulsion of the charges. The definition (4.45) is the equilibrium distribution function of the positions of charges in the Coulomb one-dimensional gas at temperature T. The total charge of the gas is $N = \int dE\,\rho(E)$, where the charge density ρ is our level density.

The relaxation of the Coulomb gas to equilibrium can be described as a random process in a time-like variable τ. During a small interval $d\tau$, the Hamiltonian matrix changes as

$$\hat{H}(\tau + d\tau) - \hat{H}(\tau) \equiv d\hat{H}(\tau) = -\frac{2N}{a^2}\,\hat{H}\,d\tau + d\hat{h}, \tag{4.47}$$

where $d\hat{h}$ is a random Gaussian force depending on the symmetry properties of the ensemble. Its average over the ensemble is defined as [compare (4.21)]

$$\overline{dh_{mn}} = 0, \quad \overline{dh_{mn}dh_{m'n'}} = \frac{1}{2\beta}\,(\delta_{mm'}\delta_{nn'} + \delta_{mn'}\delta_{nm'})d\tau. \tag{4.48}$$

All indices here run through the values given by independent matrix elements H_{mn}. The GOE evolution consists of the *drift*, supported by the effective force

$$F_{mn} = \overline{\frac{dH_{mn}}{d\tau}} = -\frac{2N}{a^2}\,H_{mn}, \tag{4.49}$$

and the *diffusion* with the corresponding coefficient

$$D_{mn;m'n'} = \frac{1}{2}\,\overline{\frac{dH_{mn}dH_{m'n'}}{d\tau}} = \overline{dh_{mn}dh_{m'n'}}. \tag{4.50}$$

According to Eq. (4.49), the individual elements H_{mn} evolve independently under the action of friction, random Langevin force, and common external oscillator field.

The resulting evolution of the distribution function $P_\beta(\hat{H},\tau)$ as the probability density is given [60] by the Fokker–Planck equation governed by the balance of the drift and diffusion processes,

$$\frac{\partial P_\beta}{\partial \tau} = \sum_{mn} \frac{\partial}{\partial H_{mn}} \left(\sum_{m'n'} \frac{\partial D_{mn;m'n'}}{\partial H_{m'n'}} - F_{mn} \right) P_\beta. \tag{4.51}$$

Let the evolution start at $\tau = 0$ with a given matrix $\hat{H}(\tau = 0) = \hat{H}_0$. The exact solution for $\tau > 0$ is

$$P_\beta(\hat{H},\tau) = \frac{\text{const}}{\sqrt{1 - e^{-4N\tau/a^2}}} \exp\left\{ -(\beta N/a^2) \frac{\text{Tr}[\hat{H} - \exp(-2N\tau/a^2)\hat{H}_0]^2}{1 - \exp(-4N\tau/a^2)} \right\}. \tag{4.52}$$

Independently of the initial value $\hat{H}_0$, the process leads to the stationary distribution (4.2) that stops the flow in the right-hand side of Eq. (4.51),

$$\left(\sum_{m'n'} \frac{\partial D_{mn;m'n'}}{\partial H_{m'n'}} - F_{mn} \right) P_\beta = 0. \tag{4.53}$$

This process determines the evolution of the energy levels E_n. Let the eigenvalues of the matrix $\hat{H}$ at the moment τ be equal to $E_n(\tau)$. At the next moment, $\tau + d\tau$, this matrix is not diagonal anymore. Here the perturbation theory is applicable shifting the eigenvalues E_n to $E_n + dE_n$,

$$dE_n(\tau) = dH_{nn}(\tau) + \sum_{m \neq n} \frac{|dH_{mn}(\tau)|^2}{E_n(\tau) - E_m(\tau)} + \cdots . \tag{4.54}$$

The higher order terms do not contribute in the linear order with respect to $d\tau$. Then the definitions leading to the Fokker–Plank equation give

$$F_n = \frac{\overline{dE_n(\tau)}}{d\tau} = -\frac{2N}{a^2} E_n + \sum_{m \neq n} \frac{1}{E_n - E_m}, \tag{4.55}$$

$$D_{mn} = \frac{1}{2} \frac{\overline{dE_m(\tau) dE_n(\tau)}}{d\tau} = \frac{1}{2\beta} \delta_{mn}. \tag{4.56}$$

In difference with (4.49), the effective force (4.55) contains, along with the oscillatory term, also the Coulomb repulsion coming from the perturbation theory (4.54). The stationary solution arises from the requirement of zero flux in the right-hand side part of the Fokker–Planck equation. The evolution of the eigenvectors again leads to the isotropic distribution at $\tau \to \infty$.

4.6. Other Features of Quantum Chaos

The level spacing distribution is the spectral characteristic the most directly reflecting the empirical data. In relatively rare cases when the experiment provides a long reliable and complete sequence of levels with identical global quantum numbers, one can also probe fluctuations of the level number within a certain energy interval. With its traditional (Dyson–Mehta) definition, one considers the sufficiently long sequence of length L of the levels belonging to the same class. This sequence is first *unfolded* [170] in such a way that the mean-level spacing along the whole sequence is 1. For the unfolded sequence, one can find the mean square deviation of the number of accounted levels from the best approximating straight line, the so-called Δ_3 statistics,

$$\Delta_3(L) = \min_{A,B} \frac{1}{L} \int_E^{E+L} dE' \, [\mathcal{N}(E') - AE' - B]^2, \tag{4.57}$$

where $\mathcal{N}(E')$ is the cumulative level number in this sequence from its start to energy E'.

A series of equally spaced levels (a regular staircase) would make $\Delta_3(L) = 1/12$. In the extreme case of a spectrum with no level repulsion, the fluctuations are far greater linearly increasing with the length of the spectrum, $\Delta_3(L) = L/15$. In the GOE, we have a disordered crystal of levels, and the quantity (4.57) grows only logarithmically,

$$\Delta_3(L) = \frac{1}{\pi^2}\left[\ln(2\pi L) + \gamma - \frac{5}{4} - \frac{\pi^2}{8}\right] = \frac{1}{\pi^2}(\ln L - 0.0678), \tag{4.58}$$

where $\gamma \approx 0.5772$ is the Euler constant. This is called the *spectral rigidity* of a random spectra. In fact, the definition (4.57) is given for a certain copy of spectra so that it is necessary to make further average going to $\overline{\Delta_3(L)}$ and taking different samples of spectra, see Ref. [170]. The Δ_3 statistics for Poisson, GOE, and GUE distributions are compared in Fig. 4.5.

Another useful measure is the level number variance $\Sigma^2(L)$ [177]. While $\Delta_3(L)$ gave us the average deviation of the spectra of given length from a regular picket fence spectrum of a harmonic oscillator, this statistics provides the variance of the level number in an interval of length L. The relation between the two statistics is given by

$$\Delta_3(L) = \frac{2}{L^4}\int_0^L dx \, \Sigma_2(x)(L^3 - 2L^2 x + x^3). \tag{4.59}$$

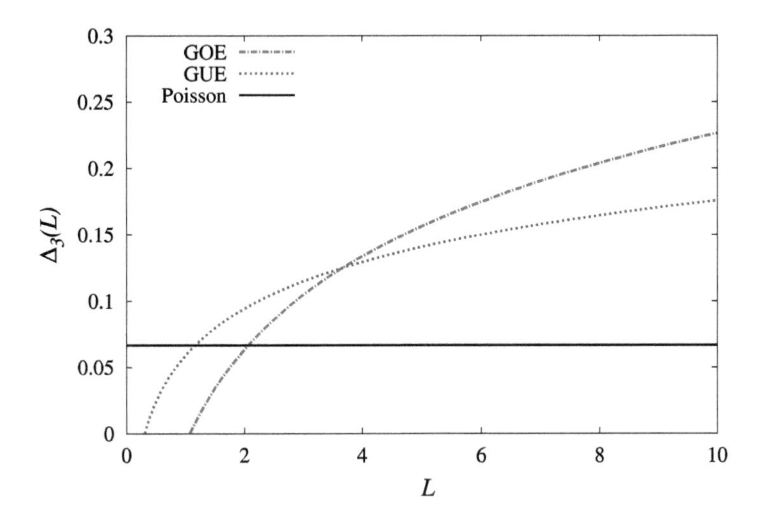

Fig. 4.5. $\Delta_3(L)$ statistics for GOE, GUE and Poisson distributions.

An interesting feature of chaotic spectra is the distribution of curvatures [140] of the wave functions as a function of a real parameter in the Hamiltonian. The averaged distribution emerges after encountering multiple avoided crossings in the process of the parameter evolution in the direction of actual interaction strength. In the chaotic regime, this distribution is significantly more smooth compared to the regular spectra. The predicted distribution,

$$P(k) = \frac{\text{const}}{(1+k^2)^{3/2}}, \quad k = \frac{1}{\pi}\frac{d^2E}{d\lambda^2}, \tag{4.60}$$

where λ is the parameter defining the strength of the chaotizing interaction, is well reproduced by the random ensemble simulations.

4.7. Level Density and Semicircle Law

There are many applications requiring the calculation of observables averaged over Gaussian ensembles. Serious mathematical problems along this road were solved, mainly by I.M. Lifshitz, Gredescul, and Pastur [150], and Mehta and Gouden [155]; these works have received an extremely high estimate by Dyson. The problem is simplified in the limit of high dimension $N \gg 1$, as we explain below (this can be applied to realistic nuclear problems).

The solution method for a long level chain can be roughly characterized as a representation of the averaged quantity in the form of a series of matrix elements H_{kl} with the ensemble averaging, term by term, and discrimination of contributions according to the parameter $1/N$. The leading principle is related to finding the terms with the maximum number of traces in the Hilbert space as every trace is proportional to $N \gg 1$. A more advanced method [234] uses the integral representations in usual and Grassmanian variables, when the expansion in inverse powers of N is derived by the stationary phase method. Below, as an illustration, we show the example (see [263] for more details) of the Green's function defined in the complex plane of the energy variable $\mathcal{E}$,

$$\hat{G}(\mathcal{E}) = \frac{1}{\mathcal{E} - \hat{H}}, \tag{4.61}$$

where, in agreement with (4.21) and (4.22),

$$\overline{H_{ij}H_{kl}} = \frac{a^2}{4N}\,(\delta_{ik}\delta_{jl} + \delta_{il}\delta_{jk}). \tag{4.62}$$

Only even orders of the expansion of (4.61) contribute to the ensemble averaging. This can be presented with the average effective *mass operator* M,

$$[\overline{G(\mathcal{E})}]^{-1} = \mathcal{E} - M(\mathcal{E}), \tag{4.63}$$

where $M(\mathcal{E})$ is given by the infinite series of diagrams, see Fig. 4.6. Here the direct line is just the inverse energy going through this line, the dotted line is the pairwise average (4.62), and the closed loop means the trace in the Hilbert space. Every crossing point corresponds to the matrix multiplication. In the first order, there is a unique loop. In the second order, we have the rainbow diagram with two concentric dotted lines and the diagram with the crossing dotted lines. The rainbow graph contains two traces while the second graph has only one trace and therefore gives a contribution of the order N^{-1} compared to the first one. We realize that in each order the main contribution comes from the rainbow graphs with non-crossing (concentric) dotted lines, Fig. 4.6(b), and therefore the maximum number of the traces. The infinite sum of the main diagrams gives

$$M(\mathcal{E}) = \frac{a^2}{4N}\,\mathrm{Tr}\,\overline{G(\mathcal{E})}. \tag{4.64}$$

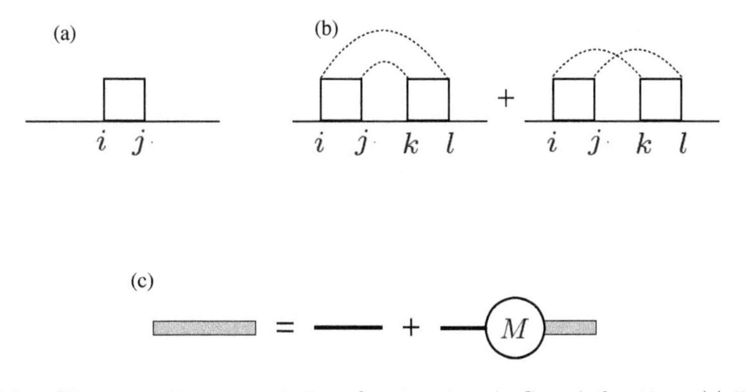

Fig. 4.6. Diagrammatic representation of contractions in Green's function: (a) Hamiltonian H_{ij}; (b) contractions in Eq. (4.62); (c) Green's function expansion through the mass operator in (4.63).

Introducing the main trace and using Eq. (4.63),

$$g(\mathcal{E}) = \frac{1}{N}\,\mathrm{Tr}\,\overline{G(\mathcal{E})} = \overline{G(\mathcal{E})}, \tag{4.65}$$

we come to the quadratic equation for this function. Asymptotically, there should be

$$g(\mathcal{E}) \Rightarrow \frac{1}{\mathcal{E}} \quad \text{at } \mathcal{E} \to \infty. \tag{4.66}$$

This determines the correct physical root of the equation,

$$g(\mathcal{E}) = \frac{2}{a^2}\,(\mathcal{E} - \sqrt{\mathcal{E}^2 - a^2}\,). \tag{4.67}$$

The level density comes as the set of the infinitely close poles, $\mathcal{E} \to E$ of the Green's function on the real axis (4.61). They are given by singularities at the limit of the imaginary part of the Green's function approaching the real axis,

$$\rho(E) = -\frac{N}{\pi}\,\mathrm{Im}\,g(\mathcal{E})_{\mathcal{E}\to E}. \tag{4.68}$$

The imaginary part of g appears only inside the interval $(-a, a)$ of the real axis which leads to the Wigner *semicircle law*,

$$\rho(E) = \frac{2N}{\pi a^2}\,\sqrt{a^2 - E^2}, \quad |E| \leq a. \tag{4.69}$$

Inside the interval, far from its ends, the level density is practically constant. The mean level spacing is $D = 2a/N$. We will see that realistic level densities in nuclear problems do not reach such a limiting semicircle form (4.67).

Developing the electrostatic analogy, the trace $Ng(\mathcal{E})$ can be interpreted as the electrostatic field created by the distribution ρ of charges in the complex plane $\mathcal{E}$. In all complex points, the charges are absent, and the electrostatic Maxwell equation coincides with the Cauchy–Riemann condition for an analytic function,

$$\frac{\partial g(\mathcal{E})}{\partial \mathcal{E}^*} = 0. \tag{4.70}$$

The electrostatic equations relate the discontinuity of the electric field on the line of charges with the charge density (4.69).

At this point, it is interesting to compare the result for the GOE to the level density in the more general Gaussian ensemble of *complex* matrices [103] that is defined by the distribution of matrix elements

$$dP(\hat{H}) \propto e^{-2(N/a^2)\,\mathrm{Tr}\,(\hat{H}^\dagger \hat{H})} \prod_{kl} d(\mathrm{Re}\, H_{kl})\, d(\mathrm{Im}\, H_{kl}). \tag{4.71}$$

The matrices here have $2N^2$ statistically independent parameters. The ensemble leads to the distribution of the complex eigenvalues z_j

$$dP(z_1, \dots z_N) = \prod_{j<k} |z_j - z_k|^2 e^{-2(N/a^2)\sum_k |z_k|^2} \prod_k d(\mathrm{Re}\, z_k)\, d(\mathrm{Im}\, \mathrm{z_k}). \tag{4.72}$$

Here we have, similarly to the GUE, quadratic repulsion of the eigenvalues in the complex plane. The result (4.72) coincides with the statistical distribution of the two-dimensional Coulomb gas in thermal equilibrium at effective temperature $T = 1/2$. The charge density at $N \gg 1$ has a constant value inside a circle of radius a,

$$\rho(z) = \frac{N}{\pi a^2}, \quad |z| < a, \tag{4.73}$$

and vanishes outside this circle.

The semicircle distribution (4.69) corresponds to the projection of the distribution (4.73) onto the real axis. As the imaginary part of the eigenvalues can have here any sign, this distribution is not appropriate for the description of decaying quantum states which are described with the help of complex energies $E - (i/2)\Gamma$ with the fixed (negative) sign of the imaginary part corresponding to the irreversible decay, $\Gamma > 0$. Later we will discuss the ensemble appropriate for that purpose.

Chapter 5

Shell Model as a Testing Ground for Quantum Chaos

5.1. Statistical Features of Realistic Calculations

When we speak about a "shell model" we have in mind the diagonalization of big Hamiltonian matrices corresponding to certain classes of nuclear systems; the corresponding modern term is *configuration interaction*. The fundamental matrix of such a study is built with a realistic set of matrix elements that supposedly accurately reflect the quantum dynamics inside the system, provide the reliable results for physical observables in reasonable agreement with available data, as well as predictions for future experiments. However, for a theorist, this approach gives much more: one can study the evolution of results under specific changes of the matrix which hopefully can unravel the underlying physics. In parallel, a huge matrix taken at its face value serves as a reasonable physical model representing the mesoscopic world even if the limitations on the orbital space make an upper part of the predicted quantum spectrum not corresponding to specific nuclear reality.

The simplest feature presenting the relation to (and the check of) the ideas of random matrix ensembles is the spectrum of energy eigenvalues. One can start with the initial ideal shell-model scheme and very weak interactions — just to remove the level degeneracy. Then the repeated diagonalizations under gradual increase of the interaction strength (with the same scale factor for all matrix elements) demonstrate the multitude of avoiding level crossings and steady evolution of the spectrum in the direction of a "disordered crystal" in each symmetry class. Taking the scale factors different for different parts of the Hamiltonian, one can also study the role of individual physical interactions in the process of chaotization.

The starting Hamiltonian matrix has, in the typically used mean-field (for example, but not necessarily, harmonic oscillator) basis, a band shape in average along the main diagonal with many empty spots being formed by the usual limitations, for example two-body interactions, and global conservation laws of angular momentum and parity (isospin can be included as well). In many practical applications, the M-basis might be computationally convenient avoiding the complicated angular momentum coupling formalism but paying the price of increasing the matrix dimension. The matrix by itself does not carry any chaotic signatures; vice versa, it looks rather regular as, with the usual limitation of pairwise interactions, the same two-body matrix elements can enter on the background of various configurations of remaining "spectator" particles. As an example we show, Fig. 5.1, the Hamiltonian matrix for the model of the ^{20}Ne nucleus with only four valence nucleons (two protons and two neutrons) in the sd-shell. There are just 46 states with total spin $J = 0$ in this space but much more states with higher spins; in the M-scheme there are 640 states with $M = 0$.

With the particle number growing, the matrix dimension increases very fast, and its *partition structure* becomes more prominent. The partition is an allowed by the Pauli principle configuration of non-interacting particles in the available mean-field space with given J^{Π} quantum numbers. Figuer 5.2 (upper part) shows the partitions 2^+ for ^{28}Si in the sd-space

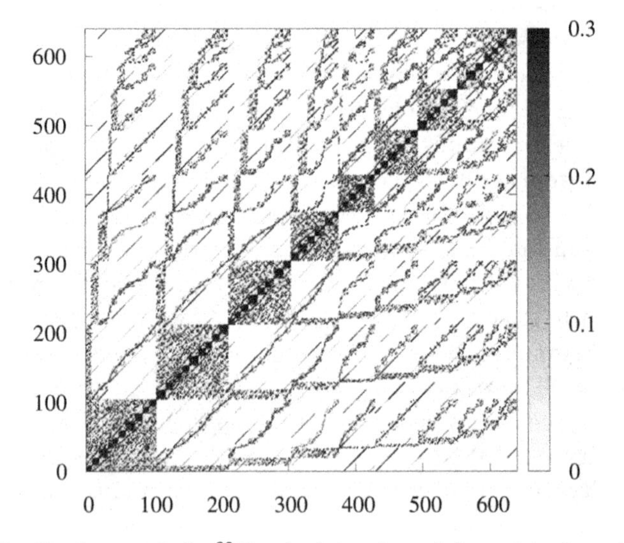

Fig. 5.1. Hamiltonian matrix for ^{20}Ne, absolute values of the matrix elements are shown on a grayscale map.

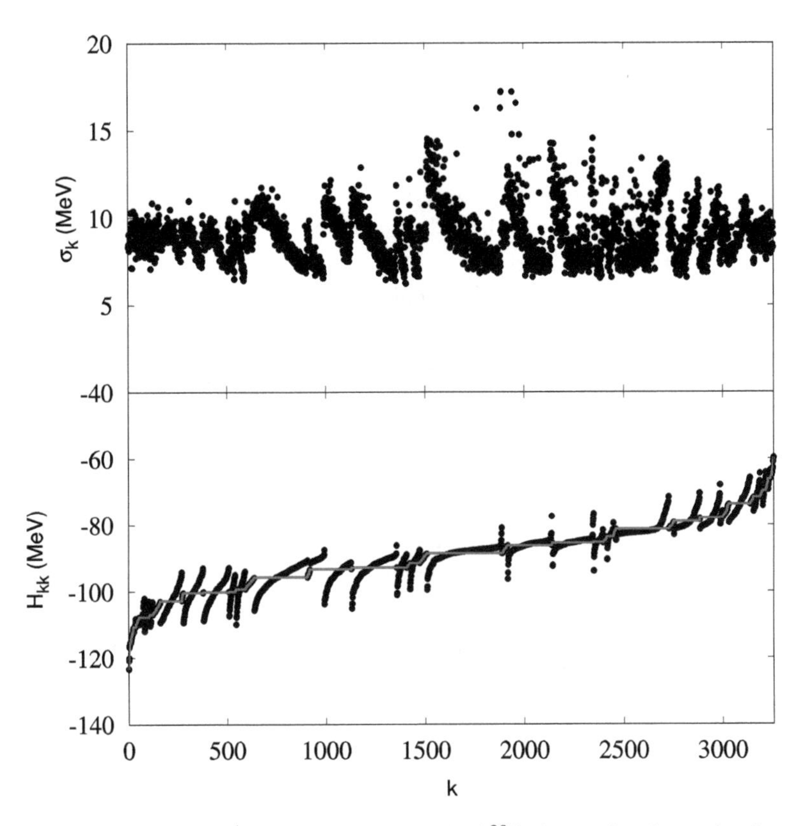

Fig. 5.2. For each basis 2^+ state in the sd-model of ^{28}Si, lower plot shows the diagonal matrix element along with the average energy within the partition (solid red line). The upper plot shows the energy dispersion of each basis state $|k\rangle$.

ordered by the diagonal matrix elements of the Hamiltonian including the interaction parts acting inside a partition. The lower part of Fig. 5.2 illustrates the centroids $\bar{E}$ of this energy for each partition [95, 189] (the traces of the diagonal part H_1 of the Hamiltonian averaged over the partition).

In a similar way, one can proceed to higher *moments* of the Hamiltonian. They can be expressed through the strength functions $F_k(E)$ of basis states (3.12),

$$\sigma_k^{(n)} = \int dE\,(E - \bar{E}_k)^n F_k(E), \tag{5.1}$$

with averaging over the states $|k\rangle$ inside a given partition. A special role is played by the second moment (see Fig. 5.2, upper panel) that is equal to the sum of the squared off-diagonal matrix elements H' originated inside a given partition,

$$\sigma_k^2 = \sum_{l(\neq k)} (H'_{lk})^2. \tag{5.2}$$

It turns out that the knowledge of these two lower moments for all partitions is practically sufficient for finding the level density for a given class of states, see below *Moments method*.

The distribution of the off-diagonal matrix elements H' is far from being Gaussian as in random matrix ensembles. In fact, the realistic distribution of the off-diagonal matrix elements in the shell-model Hamiltonian matrix is close to the Porter–Thomas distribution for nuclei and atoms [90, 260]. The semi-qualitative explanation [260] is in the significant role of multipole–multipole forces in realistic nuclear dynamics when the multipole matrix elements have a distribution close to the Gaussian. This seems to be a rather general property of atomic [90] and nuclear models including the interacting boson model [127]. Very similar features are demonstrated for random but *banded* Hermitian random matrices [57] that, in a special initial basis, are presented by a band of non-zero random matrix elements. With the band width b in the N-dimensional space, the effective scaling parameter is $\sim N(b/N)^2 = b^2/N$ [139]. In spite of similar statistical properties in the eigenbasis, we have to stress that the realistic shell-model matrices are different by selection rules and by the systematic growth along the diagonal. Therefore, the comparison may again require the unfolding procedure.

The first thing to compare with abstract random matrix ensembles is the nearest level spacing distribution. Changing the scale of all matrix elements of two-body interaction by a common factor λ we see, Fig. 5.3, that,

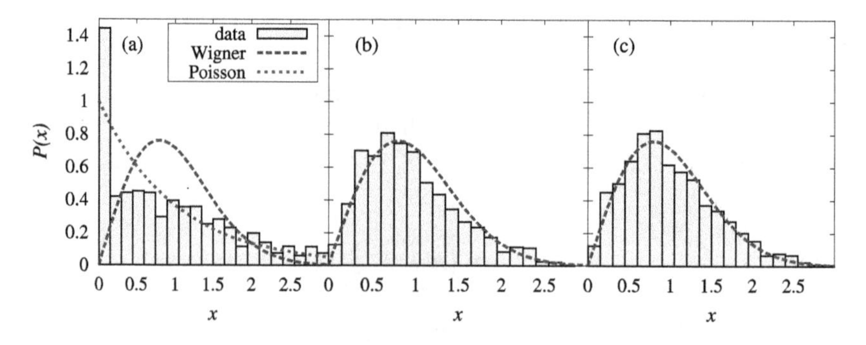

Fig. 5.3. Level spacing distribution for ^{28}Si at the interaction strength weakened consecutively by a common factor $\lambda = 0.1, 0.2, 0.3$, for panels (a), (b), and (c), respectively.

already at $\lambda = 0.2$, this distribution is quite close to the Wigner surmise (4.9). One can conclude that the level spacing distribution is the weakest signature of quantum chaos as it emerges already at a rather early stage of the chaotization process, after only few avoided level crossings. The same conclusions follow from the studies [133] of consecutive energy ratios.

5.2. Level Statistics and Dynamics

A reasonable comparison of the results in a realistic shell model to ideal properties of random matrix ensembles with necessity requires the idea of *ergodicity*: for a sufficiently high dimension N, we expect that the energy eigenvalues and properties of the wave functions from different parts of the spectrum (and even from different but similar systems) can be treated similarly to different members of a statistical ensemble. This might be correct if such an analysis excludes the lowest and the highest parts of the large spectrum where the limitations of the finite Hilbert space are especially noticeable. However, we understand immediately that the growth of complexity with increasing excitation energy that is absent in static random matrix ensembles has to be a special subject of study being a mirror of the chaotization as a process reminding the thermodynamic equilibration. Vice versa, for comparison with random matrices, this gradual evolution along the spectrum has to be eliminated as it is done with the unfolding procedures. Unfolding might be also technically necessary in order to get rich statistics.

As already mentioned, the simplest characteristic of the eigenenergy spectrum is the nearest level spacing distribution, or its later version with the spacing ratios. For the level spacings, the statistically rich analysis includes the unfolding of level energies. The function $P(s)$ makes sense only if the universal scale is used for the representatives of various parts of the actual spectrum. The results from the realistic shell-model calculations in a specific $J^{\Pi} = 2^{+}$ sectors and $T = 0$ are shown in Fig. 5.4. Here, the artificially evolving interaction strength (a common factor λ for all components of the two-body interaction) leads in ^{28}Si from the Poisson distribution to the Wigner distribution already at $\lambda \geq 0.2$. At greater values of λ, the distribution is always close to the Wigner one.

The next step of studies naturally leads from local level correlations to the long sequences of the eigenstates. For the chaotic dynamics, we expect the logarithmic growth of the spectral rigidity $\Delta_3(L)$ with the length L of unfolded levels belonging to the same class of constants of motion. In a big

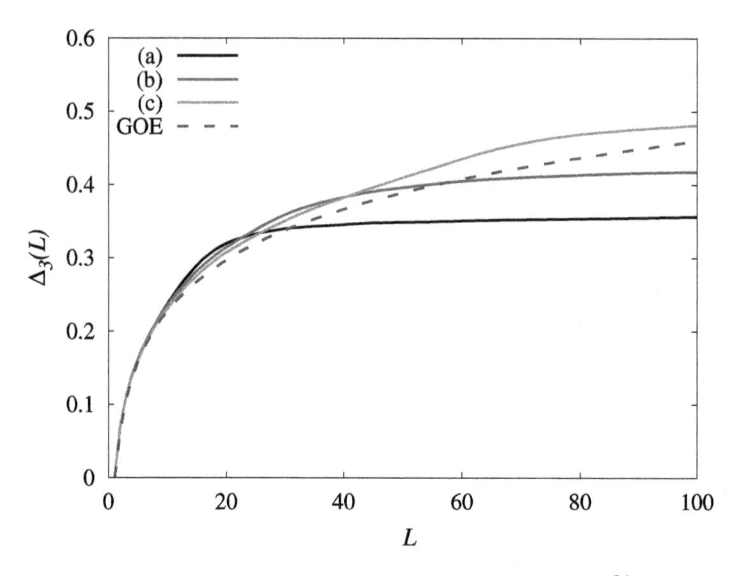

Fig. 5.4. Examples of $\Delta_3(L)$ statistics for $J = 2T = 0$ states in ^{24}Mg. Curves (a), (b) and (c) show different choices of unfolding that average density over 10, 20, and 40 levels respectively, defining the statistical corridor. The GOE result is represented by a dashed line.

matrix, it is possible to follow this distribution up to very large L where we expect deviations from the predicted GOE logarithmic behavior. Typically, the curve $\Delta_3(L)$ upbends from the GOE predictions. This behavior is known from several problems in one-body chaos [12, 253]. Experimentally this was seen in the work with a superconducting stadium billiard [111]. The long L limit is physically related to the processes occurring at short time scale where the chaotic dynamics are not pronounced yet (B. Mottelson, unpublished). Later we discuss the time dynamics in more detail but here we can simply compare this to the radioactive decay [181]. The initial stage of the time development of any nonstationary quantum state is universally determined by the unitarity of dynamics,

$$\langle k(t)|k\rangle = \langle k|e^{iHt}|k\rangle \approx 1 - \frac{t^2}{2}\left[(H^2)_{kk} - \bar{E}_k^2\right] \approx e^{-t^2\sigma_k^2/2}. \tag{5.3}$$

As discussed in [170], the $\Delta_3(L)$ statistics leads to the family of predictions forming an allowed statistical corridor, this is evident from different unfolding choices shown in Fig. 5.4. It turned out to be possible to analyze the long empirical sequences of neutron resonances for ^{235}U where the large amount of experimental data is available. This analysis allows one to conclude that few percent of levels are probably missing. In addition, with the ground-state spin 7/2 of this isotope, the compound nucleus after capturing the slow neutron can come to the spin 3 or 4 which are usually not separated by the experiment. Therefore, the observed resonances form a statistical superposition of two spin subsequences. The Δ_3 statistics indicates that the allowed values of spin indeed are present with the equilibrium statistical ratio 7/9 given by the number of projections.

A detailed study can be done for the dynamics of levels induced by the change of Hamiltonian parameters [118]. Once, with the growth of energy along the spectrum, the chaotic level structure of a disordered crystal is established, the picture of further level dynamics under some changes of parameters λ becomes generic. As described in [260], each individual pairwise level crossing is similar to a particle collision in a gas: it can be described by the term slopes $v = dE_\alpha/d\lambda$ ("velocity" of the collision), the level correlation function $\overline{E_\alpha(0)E_\alpha(\lambda)}$ (an analog to the range of forces), and the level curvature $K_\alpha = d^2E_\alpha/d\lambda^2$ (plays the role of the result of the collision as a scattering angle) [223, 252, 258].

In the limit of $N \to \infty$, the curvature distribution $P_c(K)$ is predicted [176] as

$$P_c(K_0) = \frac{\Gamma(1+\beta/2)}{\sqrt{\pi}}\,\Gamma((1+\beta)/2)\,(1+K_0^2)^{-(1+\beta/2)}, \tag{5.4}$$

where $\beta = 1, 2, 4$ for GOE, GUE, and GSE, respectively, and $K_0 = K/(\pi\bar{\rho}\bar{v^2})$ is the dimensionless curvature eliminating local variations of the slopes $\bar{v}$ of unfolded levels with density $\bar{\rho}$. As a result of level repulsion, the spectrum is "crystallized", so that the close level collisions are rare, and the probabilities of large curvatures K_0 are suppressed. The limiting behavior at large K, proportional to $K_0^{-(2+\beta)}$, directly follows from the power law $P(s) \propto s^\beta$ valid at small spacings, $s \sim 1/K$.

Figures 5.5(a) and 5.5(b) show the whole evolution as a function of the strength of residual interactions of a certain family of levels, in this case with quantum numbers $J^\Pi T = 0^+0$ in the shell-model calculations for ^{24}Mg with eight valence nucleons (four protons and four neutrons). The common strength factor changes from zero (independent particle shell model) to one (realistic calculation); no random elements are introduced. Due to the level repulsion, the whole spectrum is expanded (the gas expansion analogy). In Fig. 5.5(b) we see how several avoided level crossings lead to the "laminary" motion of levels typical to the "disordered crystal" picture. Figure 5.6 illustrates the process of establishing of the average picture of

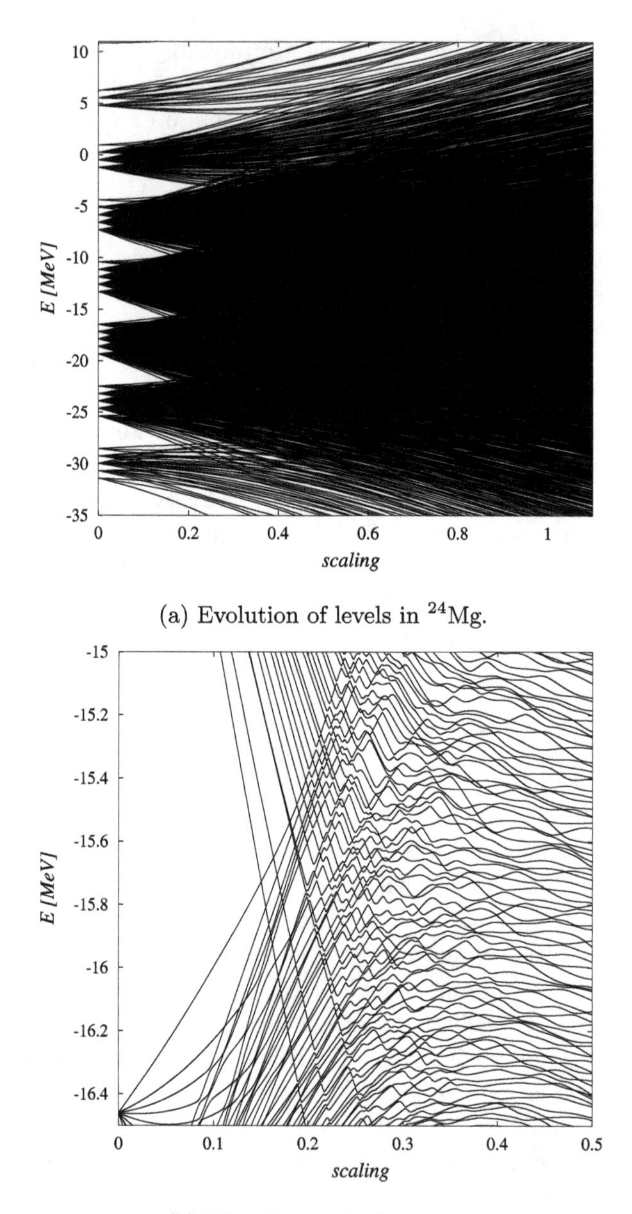

(a) Evolution of levels in ^{24}Mg.

(b) Magnified turbulent area.

Fig. 5.5. Evolution of $J = 2$, $T = 0$ levels in ^{24}Mg as a function of scaling of two-body interactions. Panel (b) shows the magnified turbulent area of multiple level crossings.

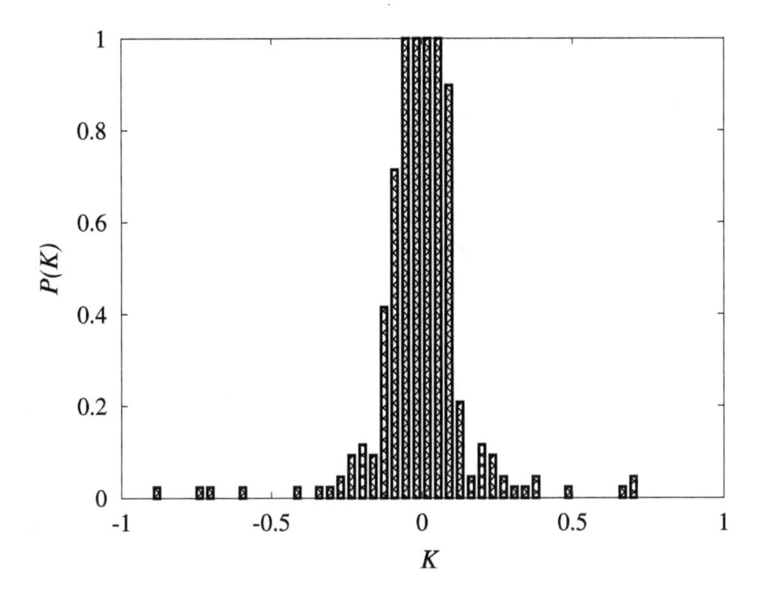

Fig. 5.6. Distribution of curvatures for the same model as in previous plots. The distribution is taken at the realistic interaction strength.

the curvature distribution. The predictions of Eq. (5.5) are fulfilled already in the strength interval 0.2–0.4. The fluctuations around this average grow when approaching the full chaoticity.

5.3. Chaotic Wave Functions

If we are interested in physics of excited states that can have observable experimental consequences, it is not sufficient to understand the energy spectrum (still, the level density is a necessary element of the practical knowledge). We have to delve into the structure of stationary wave functions and their characteristics that can be observed in actual experiments. At this point we have to distinguish the generic properties that follow just due to the extreme complexity of eigenstates from the specific features which distinguish one chaotic system from another one.

The study of the structure of individual states with necessity requires the choice of the basis. This is the point that defines the special properties of a given system. This exceptional role is naturally played by the coordinate representation in billiard-like cases [190] and by the quasienergy basis in the problems with a periodic perturbation [129]. Trivially, in the

Hamiltonian eigenbasis we do not have any complexity at all. But practically we have at our disposal only relatively simple operators capable of acting on the system, observing and quantifying its reaction, and by this response learning its specific features. In nuclear physics, the experimental techniques use external electromagnetic and strong forces acting through some multipole operators (of matter or charge distributions). To start with, we can temporary avoid specifying the basis but still assume that we have a convenient orthogonal system of states where our complicated Hamiltonian has been originally formulated using physical arguments.

Let us assume that, in the original basis $|k\rangle$, and under the condition of time-reversal invariance, the normalized stationary wave functions $|\alpha\rangle$ have typically $N \gg 1$ real random components C_k^α which are Gaussian distributed with zero average and the width $1/\sqrt{N}$. Indeed, the exact GOE distribution in the N-dimensional space,

$$P_{\rm GOE}(C) = \frac{\Gamma(N/2)}{\sqrt{\pi}}\,\Gamma(N/2-1/2)\,(1-C^2)^{(N-3)/2}\theta(1-C^2), \qquad (5.5)$$

where the θ-function puts all components into the interval $(-1,+1)$, is reduced in the required limit to the Gaussian,

$$P_N(C) = \sqrt{\frac{N}{2\pi}}\,e^{-N/(2C^2)}. \qquad (5.6)$$

If the physical observables are expressed in terms of the squares $W_k^\alpha = |C_k^\alpha|^2$ of the amplitudes C, one can use the χ^2 distribution called usually in this context the *Porter–Thomas distribution*,

$$P_N(W) = \sqrt{\frac{N}{2\pi}}\,\frac{1}{\sqrt{W}}\,e^{-NW/2}. \qquad (5.7)$$

We have to note that, contrary to the used above way leading from the GOE to the real world, in fact the Gaussian distribution (5.7) has a more general character of expressing the randomness but not being necessarily attached to the ideal GOE limit.

In realistic applications, the typical content of the basis components in a stationary state evolves along the energy axis in a rather regular way. Therefore, the characteristic dimension N_α, or some function of N_α, first of all serves as a qualitative measure of the complexity of stationary states. But, this measure being used consecutively is capable of giving also a quantitative picture as we expect it to change along the spectrum in a regular way. There are various reasonable angles of looking at such spectral characteristics.

Instead of the ensemble entropy, we can use entropy defined for individual stationary states $|\alpha\rangle$,

$$S^{\alpha} = -\sum_{k} W_k^{\alpha} \ln(W_k^{\alpha}). \tag{5.8}$$

It is critically depending on the basis $|k\rangle$ and therefore characterizes the disorder introduced by the interactions on top of the mean field taken as the basis $|k\rangle$. In the "microcanonical" limit of the uniform distribution when $W_k^{\alpha} = 1/N$, we have the absolute maximum of $S^{\alpha} = \ln N$ for all states. The states are orthogonal and, even in the GOE limit, the mean entropy is lower, $\ln(0.482N)$. Because of the logarithmic character of entropy, it might be convenient to use the corresponding length

$$l^{\alpha} = \exp(S^{\alpha}). \tag{5.9}$$

Figure 5.7 shows the information length (5.9) for all $J^{\Pi}T = 0^{+}0$ states, part (a), and for all $J^{\Pi}T = 2^{+}0$ states, part (b), in the *sd* shell model for ^{28}Si. The result is presented as a fraction of the overall dimension with 0.482 being the GOE limit. With corresponding dimensions, the ideal GOE values, constant for all states within a given class, would be 404 and 1578. In the middle of the spectra, the shell-model information length in both cases is quite close to the GOE maximum but does not reach it.

What is very important as an output of this exercise is the smoothness of the information length as a function of state energy. Comparing Figs. 5.7(a)

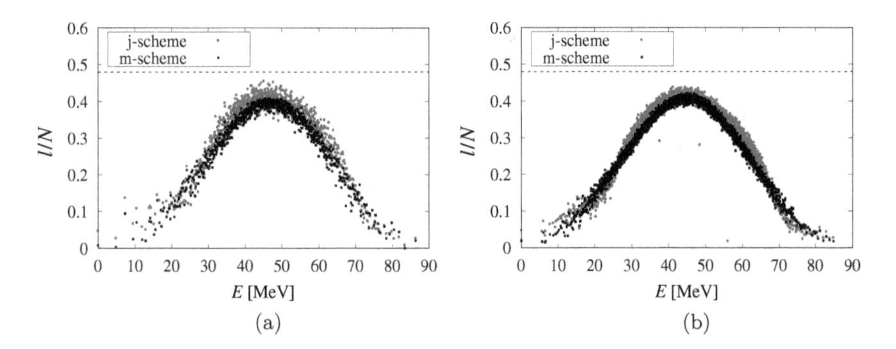

Fig. 5.7. Information length relative to corresponding dimensionality is shown as a function of excitation energy for levels in ^{28}Si. Black squares correspond to the m-scheme with a dimension equal to 93710, and red circles show the j-scheme. The dashed solid line at 0.48 shows the random matrix GOE limit. (a) 839 $J = 0, T = 0$ levels; (b) 3276 $J = 2, T = 0$ levels.

and 5.7(b), we clearly see that the fluctuations around the mean value decrease when the space dimension is increasing. All this can be naturally interpreted as the appearance of thermodynamic features. The information entropy behaves similarly to the "normal" statistical equilibrium entropy. Its smooth energy dependence makes it a *thermodynamic function*. The internal interactions equilibrate the dynamics without any external heat bath. It is possible to further study the details of this process looking at the partition structure of the Hilbert space, the isospin dependence, and partition mixture by interactions [260]. For the future, this understanding opens the way to the modified general philosophy of statistical equilibration with no heat bath.

The dimensions of the j-scheme and m-scheme are quite different. For example, 93710 m-scheme states correspond to only 839 j-scheme $J^{\Pi}T = 0^{+}0$ states in Fig. 5.7(a) but the spread of the wave functions measured by the relative information length is nearly the same in both schemes. This highlights the crucial property that while the information length is basis dependent its use as a measure of the many-body complexity relative to non-interacting particles provides a robust result. In our specific case, it is important to strictly preserve the exact rotational invariance.

Obviously, the whole process and its results are sensitive to the orbital space and the interaction details. In the review article [260], an example is shown of a different shell-model version, the set of 1183 $2^{+}0$ states in the ^{12}C nucleus including the excitations in a space of the first four oscillator shells with the Warburton-Brown interaction [247]. Here it is necessary to separate intrinsic states from the shell-model artifact of the presence of the center-of-mass excitations that breaks the partition structure. Out of 1183 general j-scheme states, 893 non-spurious internal states have to be extracted projecting out the additional center-of-mass excitations.

The distribution of the information length for intrinsic states in this shell-model example is shown in Fig. 5.8. Despite significant complexity of the non-interacting particle basis, the trends are the same. Both j- and m-schemes show low complexity for the states at the edges of the spectrum while the entropy value seems to approach the GOE values in the middle. The basis dependence of the information entropy and the different number of states being mixed in particular, result in the different l/N ratio but this does not effect the conclusions. If the intruder states with the center-of-mass excitations are artificially moved to high energy, beyond the realistic internal spectrum, one can find a corresponding branch of states with the similarly growing complexity [260].

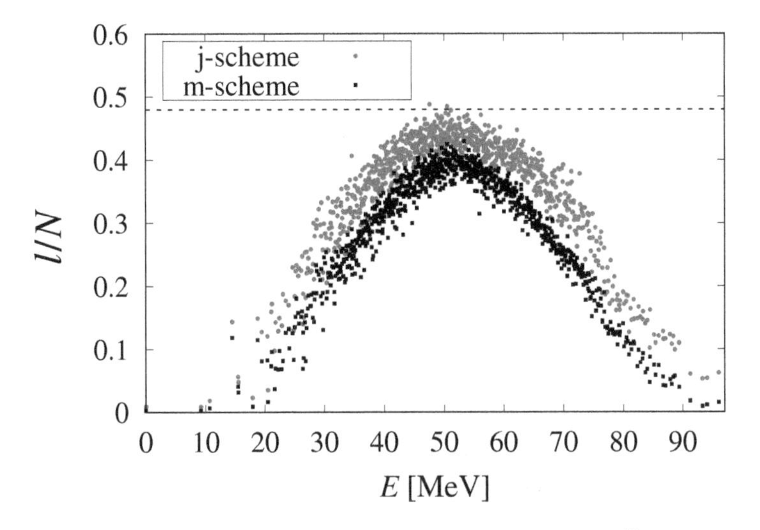

Fig. 5.8. Information length for 893 2^+ 0 non-spurious states in the ^{12}C nucleus calculated within four major oscillator shells for $(0+2)\hbar\omega$ excitations. Similar to Fig. 5.7, the j-scheme is compared with the m-scheme; the dashed solid line at 0.48 shows the random matrix GOE limit.

Typical distributions of the information length for the same shell-model examples are shown in Fig. 5.9 where the trend to the GOE values in the middle of the spectrum is clearly visible.

The complexity can be characterized by other moments of the component distribution for the stationary states, again in relation to the original basis $|k\rangle$. For example, one can take any moment of the distribution,

$$M_n^\alpha = \sum_k (W_k^\alpha)^n. \tag{5.10}$$

In many studies, the *inverse participation ratio*,

$$(\text{NPC})^\alpha = (M_2^\alpha)^{-1} = \left[\sum_k (C_k^\alpha)^4\right]^{-1}, \tag{5.11}$$

was used. This measure can be interpreted as the effective number of significant ("principal") basis component (NPC) in the stationary wave function. In the microcanonical limit of equal amplitudes, this number is just N^α, while in the GOE, because of orthogonality restrictions, it is only $N^\alpha/3$. The typical results for the realistic shell model are shown in Fig. 5.10, where we see that the NPC behaves quite similarly to the information length.

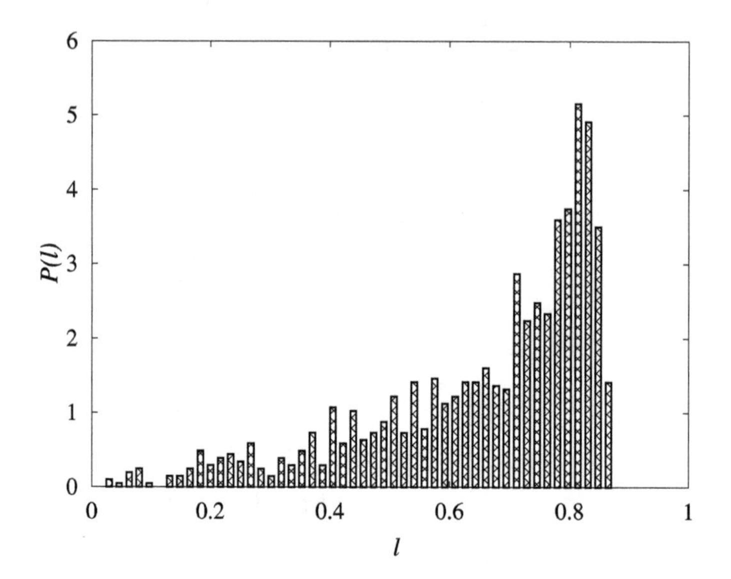

Fig. 5.9. Distribution of the information length l (relative to the GOE value) for all $J = 2$, $T = 0$ states in ^{24}Mg.

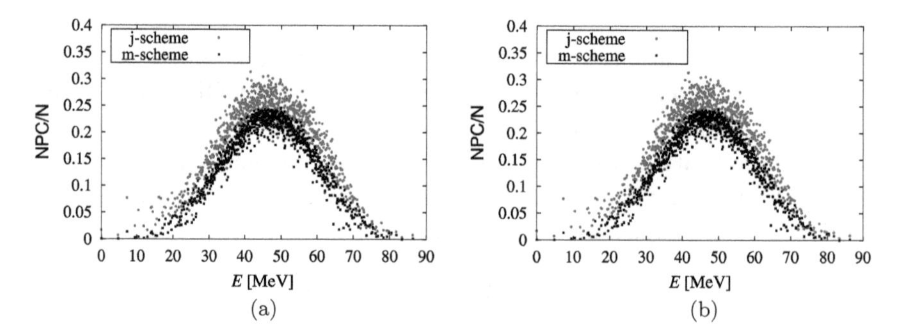

Fig. 5.10. Number of principal components for ^{28}Si states relative to the total dimensionality. Similar to Fig. 5.7, the results are shown as a function of excitation energy; black squares correspond to the m-scheme and red circles show the j-scheme. (a) 839 $J = 0, T = 0$ levels; (b) 3276 $J = 2, T = 0$ levels.

An important question is the presence of correlations between the components of a given eigenfunction. In the GOE case, the components of eigenvectors are dynamically independent. The GOE predictions emerge from the orthonormalization requirements and promise only very weak correlations, see Ref. [230]. In the limit of large N,

$$\overline{W_k^\alpha W_l^\beta} = \frac{1 + 2\delta^{\alpha\beta}\delta_{kl}}{N^2}. \tag{5.12}$$

Up to the terms of the order $1/N$, the mean value (5.12) for $k \neq l$ but $\alpha = \beta$ is equal to the product of mean values of W_k^α; if $\alpha \neq \beta$, the correlation is actually of the order $1/N^2$. In realistic nuclear physics, the dynamical correlations are definitely present — it is sufficient to recall nuclear pairing to briefly reappear later. In typical wave functions, the neighboring components (arranged by the unperturbed energy) are still correlated, mainly because they went through essentially the same road of multiple crossings. The analysis [260] shows that the decorrelation of the neighboring components proceeds on average in parallel with the increase of the informational entropy.

The discussed above measures of complexity seem to be the same as if we would be analyzing the collective states as mentioned earlier. The coherent motion creates superpositions of many simple wave functions united by a certain property, for example by the response to a certain multipole excitation. In that case, the phases of components are synchronized while here we are discussing only absolute values of mixing amplitudes. As mentioned earlier, Eq. (3.32), in order to single out the dynamically synchronized states, we can look at the phase correlator that carries over the memory of phases of the components of a stationary wave function.

The statistical analysis of the stationary wave functions reveals the stochastic character of the final ensemble of states and chaotic dynamics of the evolution process from regular to chaotic quantum states. One danger possible at this route is the change of the basis used for calculating statistical characteristics of the final set of stationary states. As discussed in [260], this process can be imitated by the regular basis transformation as in the example of a periodic chain of sites with the electron quantum states that can be described using the localized electron sites as well as the basis of waves traveling along the crystal. The expansion in terms of one basis can be extremely complicated in terms of another one, equally regular. In such cases, only basis-independent measures definitely determine the character of the dynamics.

5.4. Enhancement of Weak Effects

It seems natural to expect that, if the system is chaotic, and we have an experimental phenomenon that is unraveling at the excitation energy corresponding to quantum chaos, the systematic features of the dynamics will be suppressed or completely concealed by the extreme complexity of the wave functions in this region of the spectrum. It was already mentioned that the collective excitations appear as damping waves rather than stationary states; their spreading is described by the strength functions of the original collective states. However, this naive prognosis is not absolute. Vice versa, many specific weak but regular effects turn out to be *enhanced* at exactly such conditions.

The typical physics can be traced by the practically important occurrence of enhancement of *parity violation*. Strong and electromagnetic interactions are invariant under inversion $\hat{\mathcal{P}}$ of spatial coordinates. The Hamiltonian of the electromagnetic field contains the sum $\mathbf{E}^2 + \mathbf{B}^2$ of squared electric and magnetic fields and obviously is invariant. Recall that the electric field is a polar vector (its components change sign under spatial inversion), while the magnetic field is an axial vector (pseudovector which components are invariant under inversion). Another candidate to the electromagnetic energy would be a term $(\mathbf{E} \cdot \mathbf{B})$ but this is a pseudoscalar, changing sign under inversion. Its absence confirmed by the lack of magnetic monopoles tells us that the electromagnetic interaction preserves parity. Similarly, the Poynting vector $[\mathbf{E} \times \mathbf{B}]$ of the field momentum is a polar vector. Strong nuclear forces are also invariant under spatial inversion.

However, the weak interactions [20, 265] are really weak. Physicists have learnt about their existence observing the beta-decay. On the elementary particle level, this is manifested by the free neutron decay into proton, electron and electron antineutrino. The beta decay probability is extremely weak on the nuclear scale defining the free neutron lifetime of 15 min. Inside the nucleus, there are also the inverse processes of the proton decay or electron capture from the atomic shell transforming a proton in the nucleus into a neutron. If such processes are forbidden by energy conservation in the nucleus, in certain cases an extremely slow second order process can be observed due to the pairing effects with the double nucleon transformation and creation of two neutrinos (the search for the double neutrinoless process is still going on [71]). In principle, the existence of weak interactions gives also corrections to all nuclear processes but such effects are typically quite small, on the level of $10^{-(7-8)}$ and practically hardly observable.

What is observable, except for the very fact of a weak decay and the characteristic lifetime — a qualitative difference of weak interactions from strong and electromagnetic with respect to the fundamental symmetries of nature. Weak interactions select the "left currents". For example, the antineutrino from the neutron decay is always right-polarized (it has the right helicity, the spin projection on the momentum direction). This violation of inversion symmetry is a qualitative feature that was first observed long ago, in the famous Wu experiment [257] that had discovered the parity violation. But this is the observation of a process made possible exclusively by weak interactions. Because of the smallness of admixtures due to the weak interaction, the weak corrections to the processes driven by strong interactions were expected to be too small to be observable.

The situation changed after experiments [3, 48] on scattering of slow longitudinally polarized neutrons. If the target nucleus has the ground-state spin zero, and parity is conserved, the elastic scattering cross section of polarized neutrons was expected to be independent of the neutron helicity. The results, first being controversial, but later confirmed by several experiments, demonstrated a difference of cross sections for two polarizations on the level greater than 1% (in some cases close to 10%), Fig. 5.11. This means the enhancement of the weak interaction effects up to a factor of 10^6. The explanation [222] given almost immediately after appearance of the first data is based on simple but quite general estimates in agreement with the ideas of quantum chaos.

Consider a piece of the spectrum around the excitation energy corresponding to stationary states of a very complicated structure if measured in terms of the original mean-field basis. Let the typical number of significant basis components of a representative state be $N \gg 1$ corresponding to the strong mixing of states with exactly the same conserved quantum numbers

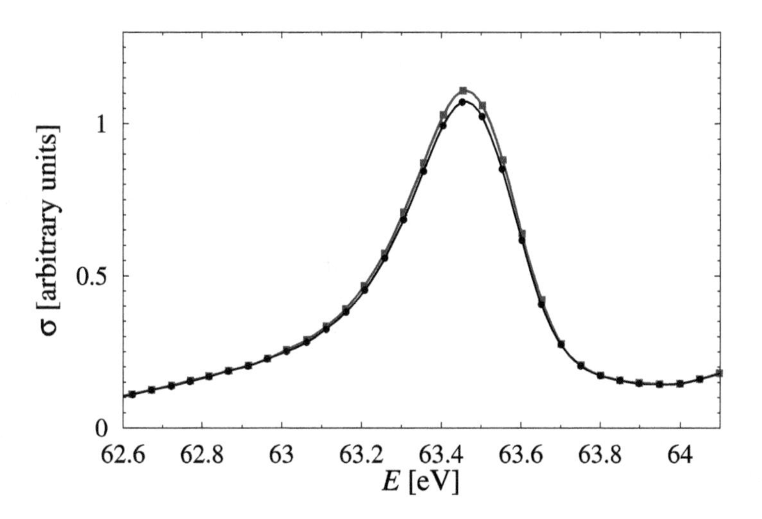

Fig. 5.11. The absorption peak in ^{238}U resonance cross section at around 63.5 eV for longitudinally polarized neutrons. Two neutron helicity states are compared, see Refs. [68, 160].

at this excitation energy. Let the weak perturbation $\hat{h}$ mix the states of two such classes: for example, the weak interactions add to a state of certain parity minuscule components of opposite parity. The perturbative admixture of a state $|\alpha'\rangle$ to the starting state $|\alpha\rangle$ is

$$\xi_{\alpha'\alpha} = \frac{h_{\alpha'\alpha}}{E_\alpha - E_{\alpha'}}. \tag{5.13}$$

If the effective operator $\hat{h}$ is of a multipole type and has simple selection rules (in the case of the weak interactions, it may be proportional to the parity changing anapole moment $\sim [\mathbf{r} \times \mathbf{s}]$ weighted with the local nuclear density), its matrix element between the stationary complicated states,

$$h_{\alpha'\alpha} = \sum_{kk'} C_{k'}^{\alpha'*} C_k^\alpha h_{k'k}, \tag{5.14}$$

can be roughly estimated as $[\nu(1/(\sqrt{N})^2]\bar{h}\sqrt{N}$. Here we assume the number of principal components $N \gg 1$ to be of the same order in both classes of mixed states, $\bar{h}$ is a typical matrix element of the perturbation between the states of these classes, ν is a number of simple states coupled by such a perturbation (usually not very big), and the last factor $\sqrt{N}$ takes care of summation with random signs of the wave function amplitudes. The energy denominators in the matrix element (5.13) are smaller by the factor of the order N compared to the energy distances between mixed simple states k, k'. As a result, we come to the *enhancement*, $\overline{h_{\alpha'\alpha}} \sim \bar{h}_{k'k}\sqrt{N}$. Of course, this qualitative estimate, the so-called *N-scaling*, can be made more precise in each specific case.

The result of this rough evaluation is the enhancement $\propto \sqrt{N}$ of simple perturbations in the chaotic region of the spectrum compared to their typical effects for simple shell-model configurations. This result is quite general being independent of the exact properties of random matrix ensembles. At excitation energy close to the neutron threshold in complex nuclei and for the realistic weak interaction, this gives the enhancement by approximately three orders of magnitude. In the case of the polarized neutron scattering, when the measured quantity η is just a difference of scattering cross sections $\sigma_\pm$ for the neutrons of different helicity by a spherically symmetric target,

$$\eta = \frac{\sigma_+ - \sigma_-}{\sigma_+ + \sigma_-}, \tag{5.15}$$

there appears an additional enhancement. Assume that the neutron is coming at the energy close to that of a p-wave resonance. Inside the nucleus, parity is violated and the p-wave is mixed by the weak interactions with a neighboring s-resonance, so that the neutron leaves the compound nucleus in the s-wave. The large ratio of the partial widths Γ_s/Γ_p in a heavy nucleus gives an additional enhancement of the order 10^3.

Finally, the total enhancement factor is about 10^6, and we expect the effect (5.15) to be revealed on the level of 10^{-2} or higher. This is exactly what is seen (by naked eye!) in the experiments, Fig. 5.11. In Refs. [68, 160] one can find the results for several other resonances. In lanthanum, the observed cross section difference was close to 10%. A special question concerns the sign of the asymmetry parameter η. It is natural to expect that this sign for various resonances on the same target nucleus has to be random. This is indeed the case for the majority of measurements. The special situation is observed for the thorium isotope ^{232}Th, where the asymmetry effect for different resonances has the same sign. This non-random signature, most probably, is related to the counterplay of quadrupole and octupole deformation specific for this nucleus [91].

The effect of chaotic enhancement is clearly seen in the experiments on the asymmetry in the fission process induced by longitudinally polarized neutrons [70, 135]. The starting polarization of a single incoming neutron induces a noticeable motion asymmetry of heavy fragments, $\mathbf{s}_n \cdot \mathbf{p}_f$, with respect to the spin of the neutron. Here there is no additional s/p enhancement but the actual asymmetry is on the level of 4×10^{-4}, Fig. 5.12, in agreement with the above estimates. An important argument in favor of the chaotic origin of the effect comes from the comparison of the asymmetry effect on observables related to the final products of the fission process. There are multiple open channels of the almost macroscopic fission process of a heavy nucleus. The products can differ by masses of the fragments, their kinetic energy distribution, etc. The asymmetry effect does not noticeably change for different final configurations. In the fission process, the fissioning nucleus goes through complicated intermediate configurations with the majority of spatial mean-field symmetries violated. During this process that is adiabatic with respect to the internal motion, the multiple crossings of many-body levels take place that leads to chaotic mixing and the enhancement of weak interaction effects. This enhancement survives the further path from the chaotic compound nucleus to final products of fission so that the asymmetry is already predetermined at the earlier chaotic stage and therefore only weakly depends on the final channels. The weak

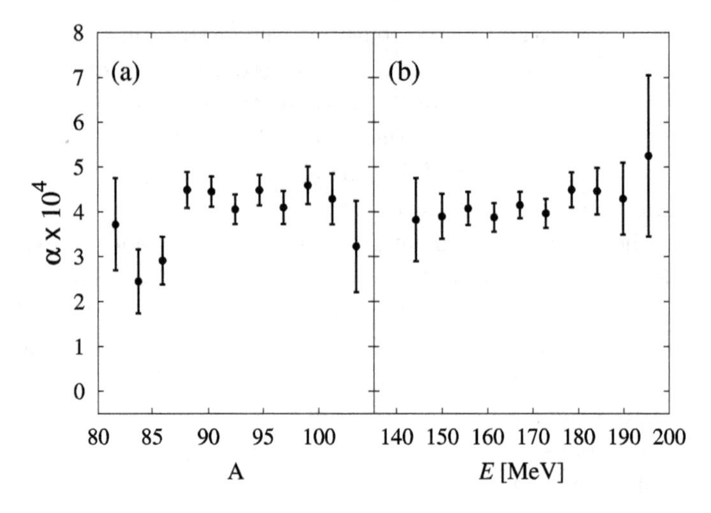

Fig. 5.12. The fragment asymmetry in various fission events with different mass (a) or kinetic energy (b) distributions, found in the ^{233}U cold neutron induced fission experiments, see Ref. [135].

parity violation is enhanced by chaotic dynamics to the macroscopically observable phenomenon.

The chaotic enhancement works in many processes which require the mixing of states of different nature at high level density, for example in radiative transitions between the superdeformed states and the states at normal deformation [114], or in the excitation of nuclear isomers [19] unrelated to fission. In the last example, the isomer can live for an extremely long time because of a very different spin or/and structure difference with lower states that forbid a gamma-radiation of low multipolarities. For artificial deexcitation, the isomer should be excited to the chaotic region where the mixing and therefore decay along the competing branch become possible. The best example is given by the rarest natural isotope ^{180}Ta with spin 9/2 and excitation energy only 75 keV that cannot undergo the radiation transition to the ground state with spin 1/2 and very different K quantum number (the projection of the angular momentum onto the intrinsic symmetry axis) so its lifetime is 10^{15} years while the ground state undergoes beta decay with the lifetime of 8 hours.

5.5. Thermalization

The main statement capable of describing the intrinsic behavior of many-body system of strongly interacting constituents is that they, without any external random elements, internally develop the chaotic behavior. This behavior can be consistently characterized in thermodynamic terms. In distinction to classical Hamiltonian systems, one can say that *quantum thermalization occurs in Hilbert space rather than in phase space* [165].

The nuclear world is traditionally described in the spirit of the Fermi liquid made of interacting nucleons. The standard scenario assumes that the deviations from the Fermi surface can be formulated in terms of quasi-particles whose lifetime close to the Fermi surface is large as the phase volume available for their decay into complex configurations is small. However, we are frequently interested in larger excitations farther from the Fermi surface. A typical nuclear example is given by neutron resonances with excitation energy 7–8 MeV mentioned in the preceding subsection in relation to the enhancement of weak perturbations. Their evolution proceeds from a simple entrance channel, a single neutron on a single-particle level embedded in the continuum, to a very complicated many-body excited state in the compound nucleus and finally to evaporation from the nucleus. After the evaporation, the nucleus can still stay in an excited state and here historically the idea of nuclear temperature appeared [249].

The nuclear temperature is not related to an external thermostat and can be measured only by the internal excitation energy and the final products of the deexcitation process. Long ago, Landau and Lifshitz in the course of *Statistical Physics* stressed that the descriptions in the quantum-mechanical language of excited states and in the language of statistical thermodynamics are essentially equivalent. Much later the same idea reappeared as the *eigenfunction thermalization hypothesis* [73, 74, 217]. Clearly, an eigenfunction capable of characterizing the general level of excitation of a medium has to be, in some sense, typical; this is especially obvious in the consideration of a region of the very high-level density where the strongly mixed individual wave functions are similar to each other. We came to the generalization of the Fermi liquid theory to higher excitations where the wave functions are chaotic. The neighboring states of the same class, after the entire process of equilibration, are getting very similar as we have seen from the evolution of the information entropy, they "look the same" [180]. This happens due to the natural process of mixing governed by the intrinsic interactions.

There are (at least) two languages for describing the state of a closed quantum system of strongly interacting constituents, either as a dense sequence of quantum many-body levels, or as a complex system in its thermal equilibrium. There are no contradictions in using those languages, they are complementary expressing coexisting properties of a complicated quantum system. In this spirit, we already introduced few definitions of the effective temperature of a closed quantum system, applying various thermometers.

One way of doing this is to rely on the informational entropy (5.8) defined for every stationary state. As this characteristic is essentially, just with small fluctuations, a smoothly changing function of excitation energy, it is natural to assume that it allows one to define the corresponding temperature which we call *thermodynamic*,

$$T_{\mathrm{t-d}} = \left(\frac{\partial \ln S}{\partial E}\right)^{-1}. \tag{5.16}$$

In agreement to information entropy, this temperature is defined for each small interval of stationary states. For the Gaussian entropy as a function of energy with the centroid E_c and width σ_E,

$$T_{\mathrm{t-d}} = \frac{\sigma_E^2}{E_c - E}, \tag{5.17}$$

with an arbitrary common normalization (the scale of the thermometer). In the middle of the spectrum, this temperature reaches infinity and then jumps to the symmetric negative region (particle-hole symmetry of the finite orbital space).

Apart from the normalization, the same result appears if one defines the effective entropy directly through the local level density $\rho(E)$ that determines the statistical weight as in the microcanonical ensemble, $\Omega(E) = \rho(E)\delta E$, by the number of close stationary states represented with equal probability in the energy interval δE. Due to the exponential growth of the level density, this is justified, maybe except for the very edges of the spectrum, leading to the microcanonical entropy $S^{\mathrm{th}}(E) = \ln \Omega(E)$. Again, the main underlying assumption is that of similarity of generic wave functions of a given class in a certain energy region, — the property of ergodicity driven by chaotic dynamics with many levels interacting and mixing. Such global definitions discard all possible remnants of random phase relationships between individual wave functions. The results are illustrated by Fig. 3.4. In every specific experiment populating the states in a given energy interval, the phase relationships between the components of the actual wave function can be different which does not influence the observables because of chaotic mixing.

Another, pure theoretical, definition of temperature can be referred to as a *quasiparticle thermometer*. As we define the stationary wave functions from the shell-model diagonalization, for each of them we can find the effective temperature $T^{\alpha}_{\text{s.p.}}$ of quasiparticles, if their occupation numbers $f^{\alpha}_{\ell j}$ can be described by the Fermi function,

$$f^{\alpha}_{lj} = \frac{1}{\exp[(\epsilon'_{\ell j} - \mu)/T^{\alpha}_{\text{s.p.}}] + 1}. \tag{5.18}$$

The actual procedure of determining the parameters is described in [260]. The behavior of all occupancies $n^{\alpha}_{\ell j} = f^{\alpha}_{\ell j}(2j+1)$ for three orbitals available in the *sd*-space is shown in Fig. 3.5 for ^{24}Mg. It is qualitatively the same in all sectors (J^{Π}) of the Hilbert space although the statistical fluctuations around the average behavior depend on the dimension of the sector. The character of the evolution of occupation numbers agrees with the ideas of a Fermi liquid. Already at the bottom of the spectrum, the interaction smears the Fermi surface. With growth of energy, the higher orbitals become systematically occupied. The thermal picture is indeed self-consistent: for example, for ^{24}Mg, with 8 nucleons above the ^{16}O core, the *sd* shell is 1/3 filled, in the middle of the spectrum ("infinite temperature"), all three orbitals are uniformly one-third occupied as expected. Similarly, in the middle of the spectrum for ^{24}Si, all three orbitals of the *sd* shell are half-filled. Here it is important that the interactions are consistent with the mean-field part of the shell model. Artificially amplified or suppressed interactions violate the consistency of the evolution process along the spectrum.

To conclude the picture, we discuss and compare various definitions of statistical quantities for a closed Fermi system of interacting particles. With no external heat bath, we can define at least three definitions of entropy: microcanonical (purely thermodynamic) function of the considered energy interval, $S^{th}(E) \sim \ln \rho(E)$, information entropy S^{α} in terms of the complexity of individual eigenstates $|\alpha\rangle$, and single-particle entropy of the Fermi liquid, also for a given eigenstate,

$$S^{\alpha}_{\text{s.p}} = -\sum_{\ell j}(2j+1)[f^{\alpha}_{\ell j} \ln f^{\alpha}_{\ell j} + (1 - f^{\alpha}_{\ell j}) \ln (1 - f^{\alpha}_{\ell j})]. \tag{5.19}$$

All these definitions, providing the temperature in a certain region of the energy spectrum in different scales, practically coincide up to fluctuations depending on the dimension of the considered sector. The microcanonical approach does not require the definition of the mean-field basis and relies only on the spectrum of stationary states of the given Hamiltonian. But all these thermometers work reliably providing the concerted results, if the demarcation between the mean field and chaotization is made consistently.

It was mentioned earlier that the experimental level density at relatively low energy can be satisfactorily described by the "constant temperature model". The parameter T of this model is not really a temperature, — this is a characteristic of the speed of growing chaotization. The exponential growth of the level density cannot continue infinitely high. At some point, this behavior is matched with the general Gaussian-like global curve of the level density. In general, at relatively low excitation above the ground state, there is no absolutely universal behavior. Here the specific features of the mean field and regular interactions define the level density including the superfluidity and collective motion. We encounter here the typical differences between even–even and odd–A nuclei, dependence on the rotational properties, etc. In the chaotic regime, the behavior becomes more standardized being determined by the statistical features of quantum chaos.

5.6. Invariant Correlational Entropy

The concept of entropy is fundamental for many branches of physics and other sciences dealing with systems which reveal a certain degree of complexity and disorder. As stressed in Ref. [175, 220], "entropy is not a single concept but rather a family of notions". In previous sections, we have seen several examples of entropy-like constructions that, being different, still reflected the same physics, the degree of complexity and the internal equilibrium reached in a mesoscopic system in the process of chaotization similar to the statistical equilibration. A short description of various modifications of the idea of entropy can be found in the introduction to Ref. [214] where one more construction, in this case invariant with respect to the choice of the quantum basis, was suggested and practically used at work.

Let the Hamiltonian $\hat{H}$ of the system contain some parameters λ which can be used to follow the process of diagonalization. In any basis $|k\rangle$ used in this process and for any stationary state $|\alpha\rangle$ (depending now on λ) we can construct a *density matrix*

$$\rho^{\alpha}_{kk'}(\lambda) = C^{\alpha}_{k}(\lambda)C^{\alpha *}_{k'}(\lambda). \tag{5.20}$$

The elements of this matrix are specific for the choice of the original representation $|k\rangle$; its trace is equal to one, and the eigenvalues are one for the state $|\alpha\rangle$ and zero for all other states. Now we make the parameter(s) λ random variables governed by the normalized probability function $P(\lambda)$ and define the average density matrix

$$\rho^{\alpha}_{kk'} = \overline{C^{\alpha}_{k}(\lambda)C^{\alpha *}_{k'}(\lambda)} = \int d\lambda\, P(\lambda)C^{\alpha}_{k}(\lambda)C^{\alpha *}_{k'}(\lambda). \tag{5.21}$$

This average matrix can also be diagonalized but its eigenvalues are not anymore 1 and 0 being real numbers between 0 and 1, while the trace is still 1. The eigenstates of this density matrix form the so-called *pointer basis* describing the influence of the random perturbation (noise) on the given state $|\alpha\rangle$ [233]. Now one can introduce the entropy of this new ensemble still specialized for a certain physical state $|\alpha\rangle$,

$$S^{\alpha} = -\,\mathrm{Tr}\,[\rho^{\alpha}\ln(\rho^{\alpha})]. \tag{5.22}$$

This entropy is invariant with respect to the choice of the starting basis $|k\rangle$; it can be called *correlational.*

Let us show a simple example [265] of how this definition works. Take a particle of spin 1/2 in a magnetic field that splits energies of the two spin z-projections on the field direction by ϵ. We apply a perturbing magnetic field $2V$ along the x-axis and a random field along the z-axis leading to the Hamiltonian

$$H = \frac{1}{2}\,(\epsilon - \lambda)\sigma_z + V\sigma_x. \tag{5.23}$$

Rotating the quantization axis to the total magnetic field, we find, at the fixed value of λ, two new states repelled to the distance Δ,

$$\sin\varphi = \frac{2V}{\Delta}, \quad \cos\varphi = \frac{\epsilon - \lambda}{\Delta}, \quad \Delta = \sqrt{(\epsilon-\lambda)^2 + 4V^2}. \tag{5.24}$$

The averaging over λ leads to

$$\overline{\cos\varphi} = c, \quad \overline{\sin\varphi} = s, \tag{5.25}$$

where now $r \equiv \sqrt{c^2 + s^2}$ is in general not equal to 1. The information entropy defined as earlier (no averaging over λ) in the basis of σ_{-z} goes to zero without a perturbation, $V = 0$, and reaches the limiting value of $\ln 2$ at strong mixing V for both split states.

This is certainly not the limit of chaos as it corresponds just to the dynamic change of the appropriate quantization axis from z to x. The situation is different if we have to average over the ensemble of noise. Then the correlational entropy (5.22) equals (for both states)

$$S = -\frac{1+r}{2}\ln\frac{1+r}{2} - \frac{1-r}{2}\ln\frac{1-r}{2}; \tag{5.26}$$

it goes to zero without averaging ($r = 1$) and to $\ln 2$ in the limit $r \to 0$ of maximum thermalization.

This elementary example shows that the correlational entropy can be used as an instrument sensitive to the processes of chaotization modeled in this definition by the parametric noise. It is well known that many phase transitions go through the stage of enhanced fluctuations [179]. This is natural if we imagine the random appearance of embryos of the new phase on the background of the old phase that becomes thermodynamically unfavorable at some value of a critical parameter. In the vicinity of this point in the parameter space, the system proceeds through the enhanced fluctuations. In the language introduced above, this leads to the noticeably enhanced correlational entropy as the value of the parameter is chaotically covering the region on both sides of the critical value.

We illustrate this point by the example [239] that allows one to evaluate the relation between pairing forces of two possible isospin value of the nucleon pair. The standard pairing with isospin projections $T_3 = \pm 1$ of the pair visibly prevails in the majority of middle and heavy nuclei. As mentioned earlier, this leads to the typical energy gap in the spectrum of even-even nuclei leading to the accumulation of states with broken pairs above this gap, renormalization of the collective moment of inertia, important effects in fission probabilities and many other phenomena in nuclear structure and reactions. There are also signatures of correlations in neutron–proton pairs with isospin $T = 0$ in nuclei [108] and nuclear matter [100, 197]. Varying the strengths of the isoscalar, $T = 0$, and isovector, $T = 1$, pairing interactions and looking at the correlational entropy of the ground state (in this example for ^{24}Mg) we see that the isovector pairing arrives much earlier than the isoscalar one. The opposite effect would require the corresponding interaction strength to be significantly stronger than in reality, Fig. 5.13.

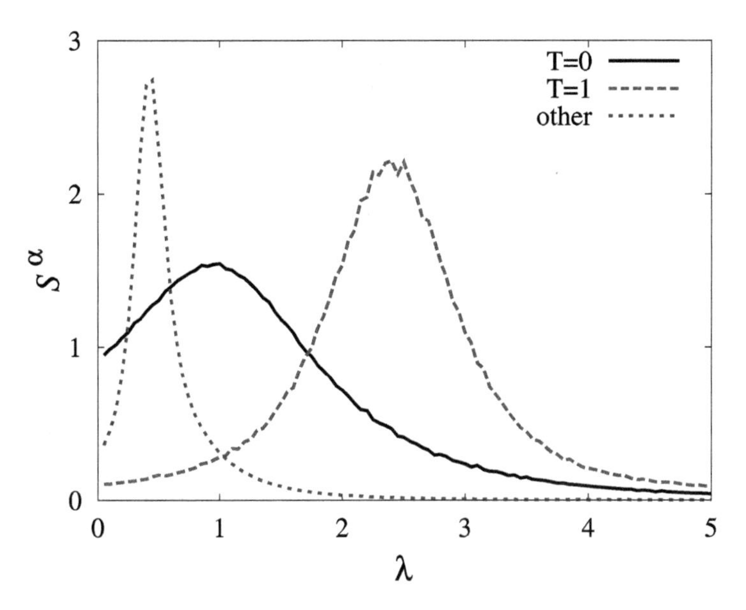

Fig. 5.13. Invariant correlational entropy (in arbitrary scale) for a ground state of ^{24}Mg shown as a function of the scaling parameter λ of pairing matrix elements for isospin $T = 0$ and $T = 1$ and of all remaining matrix elements, labeled by "other".

5.7. Level Density and Related Physics

In the previous sections we few times touched the topic of the level density in a mesoscopic system as the nucleus. Here we briefly discuss the realistic level density and some (still partly open) related problems from the viewpoint of their practical solution. In the situations when the Hamiltonian allows for exact diagonalization, we find all stationary states and therefore can directly study the level density. In fact, for practical purposes it might be sufficient to know the level density averaged over the intervals of the spectrum compared to the experimental resolution of the problem under interest. It is anyway impossible to predict the exact level positions in the region of their high density. In the problem of neutron resonances whose energies are measured quite precisely, up to fractions of eV, recall Fig. 5.11, their exact positions could be of special interest, for example in the reactor physics. In other cases, the main interest could be in the distribution functions of energies and widths of resonances. Actually, such a distinction just stresses the specificity of mesoscopic physics. The knowledge of the nuclear level density at excitation energy of the order or below 10 MeV is needed for evaluating the astrophysical reactions, especially if the medium temperature is sufficient to excite these levels with a noticeable probability.

There are several well-developed approaches to the realistic calculation of the nuclear level density explained in review papers [85, 267]. Historically, the direct calculation was based on the Fermi gas picture [39] and related but slightly more general ideas [141], including the effects of collective excitations of bosonic type [96]. For many years, the standard, and gradually improving, methods of calculating the nuclear level density were based on this old picture combined with the newest energy density functionals that allow the search for the optimal mean field enriched with the collective effects of vibrations and rotation [109, 110]. The pairing effects influential at relatively low energy can be either empirically accounted for by the energy shift [102] that mimics the spectral gap or considered by some version of the BCS theory. The experimental efforts mainly used the semi-empirical formula borrowed from the Fermi gas with the so-called "level density parameter" that in reality is usually just fit as the actual parameter of the density of states at the Fermi surface of the nucleonic gas does not work. Another complicated theoretical approach [4, 6] uses the shell model Monte Carlo method where the interaction terms are presented as one-body processes in the field of fictitious external fields which are statistically integrated out.

In spite of actual popularity and indisputable practical usefulness of such methods, their common deficiency is in the necessity to approximate the actual nuclear interactions by their simplest terms, essentially only of collective origin. The multiple interaction processes of incoherent collisional character are neglected. In the Monte Carlo methods, this is practically necessary due to the notorious *sign problem* leading to strong instability of calculations. As a result, in many examples, the outcome for the level density is a rather irregular function of excitation energy reflecting the substructure of single-particle orbitals. In fact, usually neglected additional two-body matrix elements of non-collective character would add to the width of the level density and make it a more smooth function.

Now we briefly describe the computational process, the so-called *moments method*, that uses the full shell model Hamiltonian taking into account all interaction terms and not requiring the full diagonalization of the huge matrix. The weak point of this approach is that it is necessary to work in the limited orbital space and therefore rely on the effective

Hamiltonian in this space where the presence of the shells not included explicitly is only partly compensated by phenomenological renormalization of matrix elements. Currently, such a program is fully developed for *sd*- and $pfg_{9/2}$-shells. In the cases when the full numerical diagonalization is practically possible, the results for relatively low energies (including astrophysically interesting cases) are almost identical to the exact ones.

Instead of the full numerical diagonalization, the moments method uses statistical arguments based on the broad experience with mesoscopic complexity. The mathematical foundation of this approach is in the *statistical spectroscopy* developed in the works of the French school [95] and explained in the book by Wong [255]. The recent formulation of the method was a result of several consecutive improvements [122–124, 200]. First, we need to have at our disposal a reliable shell-model Hamiltonian in a certain orbital space presented as a sum of the mean-field part $\hat{H}_0$ and interaction part; until now only two-body forces were practically employed but the use of many-body contributions is also possible.

Let $\rho(E;\alpha)$ be the level density as a function of energy and exact quantum numbers α defining the class of states. The total number of states did not change after switching the interaction on. As stated earlier, each unperturbed partition p received many admixtures in the process of this evolution. This process can be described by a certain propagator (Green's function) $G_{\alpha;p}(E)$ that transforms the non-interacting configurations into realistic eigenstates at energy E. With the dimension $D_{\alpha p}$ of the states of given class in the partition, we look for the final result of multiple avoided level crossings,

$$\rho(E;\alpha) = \sum_{p} D_{\alpha p} G_{\alpha;p}(E), \tag{5.27}$$

with the sum over all partitions having states with the given quantum numbers α. Based on the statistical spectroscopy, it is assumed that the Green's function is essentially similar to the Gaussian $\tilde{G}(x,\sigma)$

$$G_{\alpha;p}(E) = \tilde{G}(x = E - E_{\text{g.s.}} - E_{\alpha p}; \sigma_{\alpha p}). \tag{5.28}$$

The Gaussian with a tilde is the so-called *finite range Gaussian*: it has the wings of the standard Gaussian cut at $|x| = \eta\sigma$ with η a parameter that is selected empirically but always close to 2.8, see the specific discussion in [200]. This cut-off takes away the unphysical tails as the level density is not Gaussian at the actual edges of the spectrum. The ground-state energy is another parameter that can be found by the shell-model calculation, or by one of the empirical methods. Then the new Gaussian has to be normalized.

Apart from these parameters which in practice can be taken universal, the procedure of further calculation is straightforward. The Gaussian is completely defined by two quantities, its centroid and the width. They are taken from the first two moments of the Hamiltonian without necessity to diagonalize a huge matrix,

$$E_{\alpha p} = (1/D_{\alpha p})\,\mathrm{Tr}^{\alpha p} H, \tag{5.29}$$

$$\sigma^2_{\alpha p} = (1/D_{\alpha p})\,\mathrm{Tr}^{\alpha p}(H^2) - E^2_{\alpha p}. \tag{5.30}$$

In fact, both moments can be found just from the Hamiltonian matrix (diagonal and off-diagonal elements, respectively), when it is technically simpler to use the M-basis of states with a certain quantized angular momentum projection avoiding the complicated algebra of $3nj$-symbols. The difference of the results for M and $M+1$ gives the contribution of $J = M$. In a special case when the configuration space contains several oscillator shells of

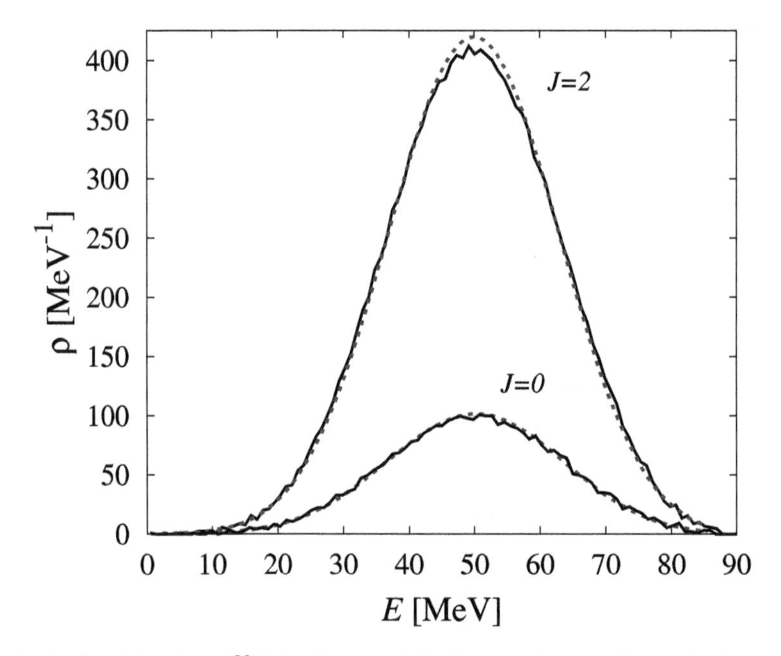

Fig. 5.14. Level density of ^{28}Si for $J = 0$ and $J = 2$ states from shell-model calculations within the sd-valence space. Exact results from the full shell model diagonalization are compared with the level density found by the moments method.

various parities, the shell-model approach contains fictitious states emerging from the center-of-mass excitation. These states are to be excluded, for example with the help of recurrence relations [124].

Actual algorithms, with possibility of parallelization, are explained and made available in Ref. [202]. Figure 5.15 from [203] shows the agreement of the moments calculation with the total shell-model diagonalization. The small differences at the maximum of the curve show that the fourth moment would be required to get the absolute agreement. However, this maximum is, as a rule, outside of the applicability region of the shell-model calculation made in the restricted space and without continuum effects. The two-moment approximation works quite well in all cases when the full shell-model results are available for comparison. The approximation of the exact Green's function by its two lowest (and renormalized) moments gives practically the exact result for empirically needed applications, except the region of the lowest energy in the immediate vicinity of the ground state, where one can use either other methods or the empirical constant temperature model that, as discussed earlier, should be matched with generic Gaussians at higher excitation energy.

The results of the moments method for the nuclei with a reliable shell-model Hamiltonian, for example [132, 266], confirm the constant temperature model (3.7) for the energy behavior of the level density at relatively low excitation energy. It follows from these broad calculations that the effective temperature is not related to the pairing interaction and practically does not change if this interaction is artificially excluded. This temperature parameter is rather related to the speed of equilibration in the system, its approach to the chaotic stage. As such, it shows characteristic changes from nucleus to nucleus. For example, in the *sd*-model, the fastest chaotization is revealed by the ^{28}Si nucleus (half-filled shell space) where all components of spin- and isospin-dependent interactions are at work. An interesting problem for future studies would be the theoretical prediction of the actual value of this parameter.

5.8. Time Dynamics

For a long time, the quantum chaos was considered by researchers from the viewpoint still reflected in the current Wikipedia definition: *Quantum chaos is a branch of physics which studies how chaotic classical dynamic systems can be described in terms of quantum theory. The question the quantum chaos seeks to answer is: What is relationship between quantum mechanics and classical chaos?* The statement of priority of classical mechanics sounds not reliable. We know now quite well that classical behavior comes as a limiting case of quantum theory under conditions of processes with action much greater than the quantum $\hbar$. The famous instability of classical orbits in chaotic dynamics described by the Lyapunov exponents [79] is the property of the trajectory singled out by the classical least action principle at the strong sensitivity to initial conditions. In quantum chaotic dynamics of an isolated system we have the dense energy spectrum of very complicated, orthogonal, but physically similar stationary states with analogous observable properties. Following quantum dynamics in such a system by means of simple observables we do not expect the Lyapunov exponential instability as multiple neighboring states (or trajectories in the path integral) have analogous properties.

In this subsection, we show the direct calculation [246] confirming such ideas. Assume that the system under study is excited by a measurement that provides a certain value of a selected classical observable. The initial stage, far from corresponding to the exact stationary state but defined with an exact starting point, relaxes with time into an equilibrated superposition of stationary states, with no exponential instability. The law of relaxation is practically universal being similar to what we could expect for a random matrix ensemble. This rather confirms the definition by Chirikov (1992): *Quantum chaos is finite-time statistical relaxation in discrete spectrum* [66].

Our object is a closed quantum system with the well-defined Hermitian Hamiltonian without any random elements. The time development of a *regular* quantum system can be illustrated by a primitive example of a coherent state of a harmonic oscillator that is an eigenstate $|\alpha\rangle$ of the annihilation operator,

$$\hat{a}|\alpha\rangle = \alpha|\alpha\rangle. \tag{5.31}$$

The time dependence of such an initial state is a phase rotation, when the expectation values of the coordinate $Q(t)$ and conjugate momentum $P(t)$ are rotating according to $(m = \hbar = 1)$

$$Q^{(\alpha)}(t) = \sqrt{\frac{2}{\omega}}\,\mathrm{Re}\,(\alpha e^{-i\omega t}), \quad P^{(\alpha)}(t) = \sqrt{2\omega}\,\mathrm{Im}\,(\alpha e^{-i\omega t}). \tag{5.32}$$

The survival probability trivially oscillates periodically returning to the initial value of 1,

$$\langle\alpha(t)|\alpha(0)\rangle = \exp[-4\alpha^2 \sin^2(\omega t/2)]. \tag{5.33}$$

Very different dynamics are seen in the case of the behavior corresponding to the random matrix ensembles. We do not limit ourselves here by precisely one of the canonical matrix ensembles but just assume a general disordered system in the following meaning. Let $|\alpha\rangle$ be the stationary states of a closed system, while the complete orthonormal basis of states $|k\rangle$ is selected arbitrarily, for example as eigenstates of a Hermitian operator $\hat{Q}$ that is not a constant of motion. Then we can assume that, in average,

$$\overline{\langle k|\alpha\rangle} = 0, \quad \overline{\langle k|\alpha\rangle\langle\alpha'|k'\rangle} = \frac{1}{N}\delta_{\alpha\alpha'}\delta_{kk'}. \tag{5.34}$$

Here $N \gg 1$, maybe slightly different for different non-conserving operators forming the basis $|k\rangle$, is the measure of the effective dimension. The time development of the system can be described by the propagator matrix

$$G_{kk'}(t) = \langle k|e^{-i\hat{H}t}|k'\rangle. \tag{5.35}$$

Under assumption of the limiting behavior (5.34), the ensemble-averaged propagator is universal,

$$\overline{G_{kk'}(t)} = \sum_{\alpha} e^{-iE_\alpha t}\delta_{kk'} \equiv G(t)\delta_{kk'}. \tag{5.36}$$

In the random limit, the function $G(t)$ is universal being the same for all initial states up to a manifold of measure zero at the edges of the large matrix. Starting with a certain typical state $|k\rangle$, we populate various states $|k'\rangle$ in general different from $|k\rangle$, in average with comparable probabilities. Taking initially the eigenstate $|k\rangle$ of $\hat{Q}$, we define the evolution of the expectation value of this operator,

$$Q_k \Rightarrow Q_k(t) = \sum_{k'} Q_{k'}G_{kk'}(t) = Q_k|G(t)|^2 + \frac{1-|G(t)|^2}{N}\sum_{k'\neq k} Q_{k'}. \tag{5.37}$$

When, having in mind a multipole moment with the zero full trace, or just normalizing the operator $\hat{Q}$ in this way, we see that the second term in Eq. (5.37) typically is of order $1/N$ compared to the first one. Therefore, in the main approximation, the time evolution of the observable from the initial state with its certain value is *universal*,

$$Q(t) = Q(0)|G(t)|^2. \tag{5.38}$$

The calculations can be performed analytically for the limiting GOE case of chaotic dynamics. In the following numerical examples, the matrix dimension is 10^4. The ensemble level density is given by a semicircle of radius $R = 2$. The energy scale is determined by the variance $R/2$. The propagation kernel that is actually the Fourier transform of the level density is given by the Bessel function J_1,

$$G(t) = \int_{-2}^{2} \frac{d\epsilon}{2\pi} \sqrt{4 - \epsilon^2}\, e^{-i\epsilon t} = \frac{1}{t} J_1(2t). \tag{5.39}$$

For the conjugate momentum we obtain, in terms of the derivative G',

$$P(t) = \frac{dQ(t)}{dt} = Q(0)(G^* G' + G'^* G), \tag{5.40}$$

where, in the GOE limit, we use

$$G'(t) = -\frac{2}{t} J_2(2t). \tag{5.41}$$

Initially, $P(0) = 0$ because the initial decay rate of unstable states is always zero as in the standard exponential radioactive decay [181]. In such calculations, it is convenient to use the dimensionless time, $t \to \lambda t$, where λ is related to the characteristic energy of the spectrum (the level density radius in the GOE case).

Figure 5.15 shows the evolution $Q(t)$ for several initial eigenstates of an operator $\hat{Q}$ with different values $Q(0)$. The dashed lines show the analytical result (5.39) that is nearly indistinguishable from the numerics. The

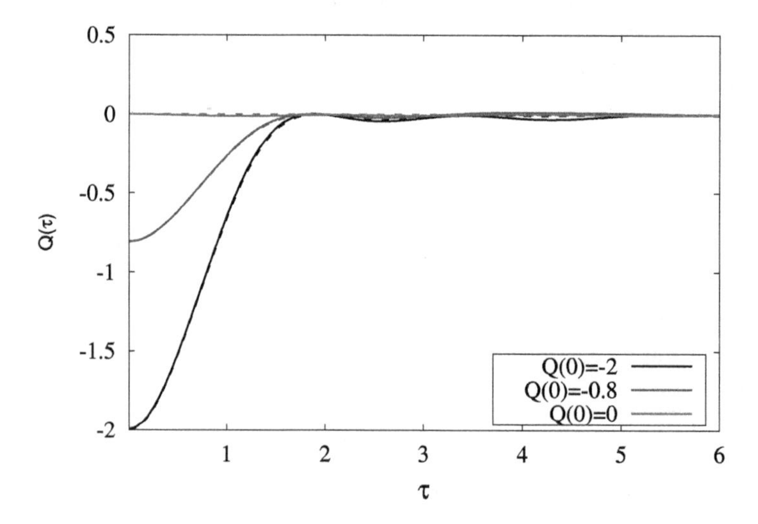

Fig. 5.15. Time evolution of the observable variable $Q(\tau)$, where $\tau = \lambda t$. Numerical studies show full agreement with the analytical results.

evolution consists of the prethermalization dependent on the choice of the initial state following by the universal equilibrated (thermalized) stage with almost identical small fluctuations. The same, practically exact, agreement with the above predictions takes place for the momentum and for the fluctuations of coordinates and momenta. In all numerical studies, a single realization of the GOE is used with no averaging, and the quality of agreement shows that the statistical assumptions are valid for each individual realization. The agreement improves for higher space dimensions. The behavior of the number of principal components, with its different definitions, is discussed in more detail in Ref. [246]. Typically, it starts in agreement with the normal behavior of the decaying state but after sometime acquires the oscillating stage with maxima at the expected values for the chaotic wave functions and minima (zeros). This is a reflection of the finite space with the typical Weisskopf time $\sim \hbar/D$ corresponding to the long-time periodic dependence on interaction resolving a quantum grid of states with the mean energy spacing D.

Now we can look at the situation in a real mesoscopic system of strongly interacting particles [246] with reliably described dynamics. The extreme assumption in the spirit of canonical random ensembles that the matrix elements of the interaction between any two many-body states are statistically identical so that the system is fully invariant under orthogonal basis transformations is certainly too far-reaching. We know that the realistic many-body systems usually have the mean-field basis that is special and thus violates orthogonal invariance. Moreover, with the standard two-body interaction, instead of the full randomness, the same elements appear repeatedly in the Hamiltonian matrix being identical on different backgrounds of the spectator particles. However, the experience partly reflected in previous sections tells us that many of the results discussed above for random matrices will remain approximately valid. The survival probability for each individual state can be different, especially for the states at the edges of the spectrum or in the case of a relatively weak interaction. It is seen quite well in the *interacting boson model* [127] that the survival probability of initial states is essentially universal up to some moment when we have specific fluctuations (an analog of thermal equilibrium) on a low probability level.

The time dynamics for a realistic nuclear example was considered in [246]. The *sd*-shell model for ^{24}Mg is known to describe well low-lying characteristics (energy spectrum, transition rates, etc). In spite of the presence of only eight valence particles (above the fixed core of ^{16}O), the dynamics realistically predict observed single-particle as well as collective modes and rotational bands. At some excitation energy, the spectrum becomes very dense and the stationary wave functions are expected to have chaotic properties. A non-stationary initially excited state in the process of quantum relaxation acquires multiple admixtures of those complicated states. Limiting ourselves by the discrete energy spectrum we are looking at time intervals of statistical equilibration which are still smaller than the lifetime with respect to the irreversible decay into continuum.

Attempts to theoretically describe the time evolution of a quantum many-body system and formation of a compound nucleus are frequently based on step-wise sequences of transitions between different classes of excited states distinguished by a number of particle-hole excitations [2] counted from the non-interacting ground state, see also [44, 45, 228] and references therein. Such processes traditionally can be described by a tree-branching population dynamics as in a real chain of radioactive transformations. Below we give an example of the exact solution of time-dependent quantum dynamics avoiding assumptions of the tree process where the motion has been essentially assumed to proceed only in the forward direction due to the higher dimension of every next class of states. We will see that, in a realistic chaotic many-body dynamics already started from the non-stationary state with all interactions present, this assumption does not work (except maybe for special models like spin systems with interaction of close neighbors). More probably, the system promptly enters the region where many states with close energy have a similar population and the thermalized motion does not have a certain direction (the time arrow is practically absent).

In the realistic example of the nuclear *sd*-shell model for ^{24}Mg with four neutrons and four protons (isospin $T = 0$) occupying $1d_{5/2}, 2s_{1/2}$, and $1d_{3/2}$ single-particle orbitals on top of the inert core of ^{16}O, we use the popular two-body USDB interaction [55]. Our variable $\hat{Q}$ will be the nuclear mass quadrupole moment Q_{20}, the tensor component with angular momentum 2, its projection $M = 0$ and isospin zero. Starting with the eigenstate of this operator, we guarantee that the dynamics will be confined to the subspace with quantum numbers $JMT = 200$ and positive parity; the dimension of this subspace in the model is 1206. The quadrupole moment of the actual first stationary state $J^{\Pi} = 2^{+}$ is $Q_0 = -1.853$ indicating significant collectivity at low energy. In Fig. 5.16, the quadrupole moments of all stationary 2^{+} states in ^{24}Mg calculated in this shell model are shown.

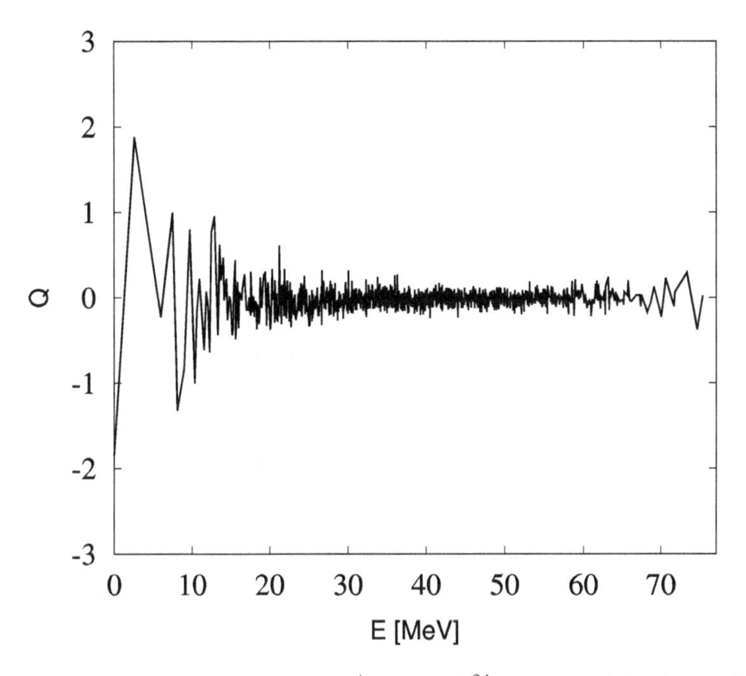

Fig. 5.16. Quadrupole moments of all 2^+ states of ^{24}Mg in the full solution for the *sd*-model space shown as a function of excitation energy.

We start with the state $|\alpha^\circ\rangle$ emerging immediately after the measurement at $t = 0$ of the quadrupole moment that provided, supposedly with high precision, a value Q°,

$$\hat{Q}|\alpha^\circ\rangle = Q^\circ|\alpha^\circ\rangle. \tag{5.42}$$

Starting with this initial state we look at its quantum evolution for $t > 0$,

$$|\alpha^\circ(t)\rangle = e^{-i\hat{H}t}|a^\circ\rangle. \tag{5.43}$$

The state $|\alpha^\circ(t > 0)\rangle$ is not an eigenstate of the quadrupole operator anymore. The survival probability for different initial states $|\alpha^\circ\rangle$ is shown in Fig. 5.17. Apart from long-time tails, the curves show remarkable identity reflecting similarity in the strength functions of the starting states. This uniformity of the relaxation process comes after rescaling the time with the help of appropriate dimensionless units $\lambda_a t$, where λ_a is the energy uncertainty in the initial state. In the same way, one can find the time dependence of the number of principal components describing the expansion of the state structure with time. In the same approximation as in

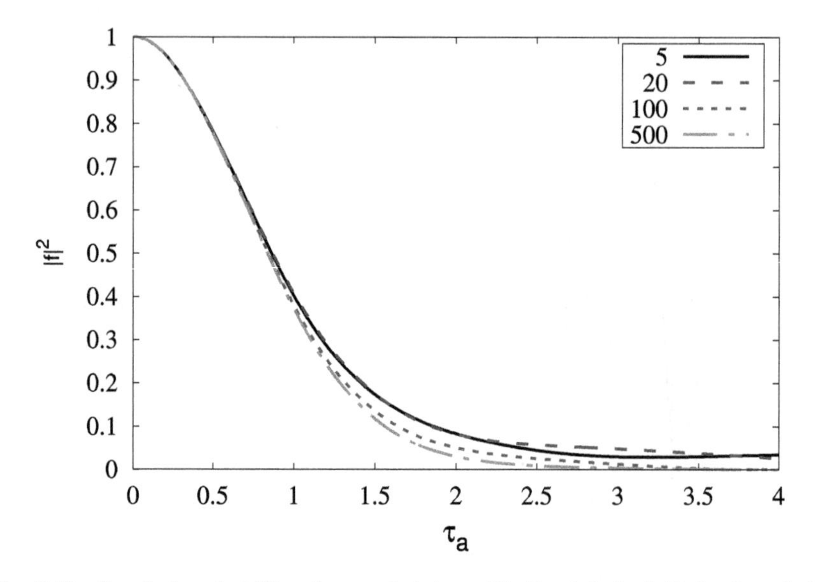

Fig. 5.17. Survival probability of several states with the definite initial value of the quadrupole moment. The initial states $\alpha^\circ = 0, 1, 2 \ldots$ are labeled sequentially in order of the increasing eigenvalue of the quadrupole moment Q° in Eq. (5.42); the states 5, 20, 100, and 500 are shown. See Ref. [246] for details.

(5.41), the chaotic dynamics predicts (in the scaled units) the dependence t^6 up to a limit of saturation on the level of the order N, where only small fluctuations survive. This is indeed what we see in the exact calculations for the shell model, Fig 5.18.

The results are sensitive to small violations of fundamental symmetries as shown in [246] by a tiny non-conservation of isospin that leads to the sharp increase of the accessible dimension. This asymmetry becomes clearly visible after initial time of "normal" isospin-invariant propagation. Few states with the largest quadrupole moment of collective origin retain a noticeable mean quadrupole moment even after a long evolution. This is a typical behavior of collective modes, like giant resonances in nuclei, which are not stationary states but doorways to mixing (damping) with numerous compound states [16, 210, 213]. Nevertheless, the collective strength being spread over many underlying stationary states still is observable being concentrated in a certain spectral region. This is a quantum analog of *scars* known in classical chaotic systems [131]. Many interesting details of the behavior of the selected observable, in this example $Q(t)$, are discussed in [246].

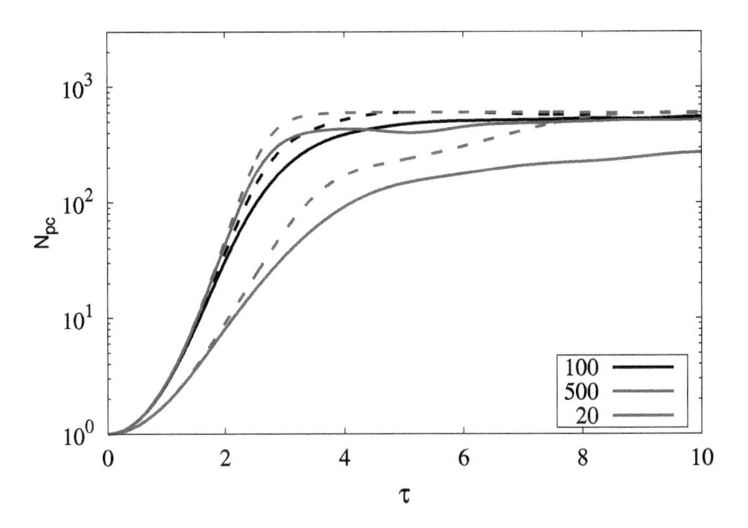

Fig. 5.18. Number of principal components as a function of time for several initial states. Dashed lines show an approximate result. The initial states with $\alpha^\circ = 20, 100$, and 500 are shown.

To conclude this rich topic, we formulate main results. The extreme chaotic limit represented by the GOE provides asymptotically exact analytic expressions describing the universal time evolution of a generic initial state, its spread in the Hilbert space and the corresponding number of principal components, a quantum counterpart for the classical propagation in the phase space. In the limit of orthogonal invariance, the time evolution is defined exclusively by the survival probability of the initial wave function. In realistic quantum many-body systems, the survival probability remains an important basis-independent generic characteristic of complex non-stationary wave functions. Further, realistic systems with the skeleton of a regular mean field show an extended decay of perturbations, much longer than the timescale given by the survival probability. This dynamics typically cannot be described by the tree branching with several intermediate stages that involve every time a new class of states (although such models are also possible). The end of the characteristic time evolution allows, along with the irreversible decay to continuum as in the radioactive decay, just thermal-type fluctuations on the level determined by the mean energy stored in the initial non-equilibrium state.

The realistic quantum dynamics of a mesoscopic system does not produce any quantum manifestation of the classical Lyapunov exponent. Classical extreme sensitivity to initial conditions is smeared by the quantum-mechanical uncertainty. Using the language of the Feynman path integrals, in a realistic many-body system, the slightly modified paths that exponentially diverge in the classical regime have nearly identical actions and become smoothly averaged in quantum mechanics. This claim refers to finite systems where, in contrast to their macroscopic limit, the interaction does not immediately induce phase transformations as in the case of Fermi-gas unstable with respect to Cooper pairing.

Chapter 6

Nucleus as an Open System

6.1. Briefly About Continuum

Strictly speaking, all excited nuclear states (and, frequently, even the ground state) have a finite lifetime and therefore belong to the continuum spectrum. However, for gamma radiation, weak and alpha (and cluster) decays, or spontaneous fission, the lifetime might be quite long by the nuclear scale so that the energy of such states can be considered as being exactly defined (very small energy width). Our shell-model exercises are fully applicable to such situations. If we consider states above the particle decay threshold, the energy spectrum is continuous, the widths of resonances become significantly greater, and new phenomena related to the structure of the nucleon continuum enter the game. With further growth of excitation energy, the widths of neighboring resonance states start to overlap that creates a completely different physical situation.

Many nuclei (not too close to drip lines) have binding energy approximately 6–8 MeV per nucleon. This energy can be brought to the nucleus through any nuclear reaction that uses an appropriate projectile. A typical "soft" process is the absorption by a nucleus of a very slow ("thermal" in reactor terminology) neutron. Any type of reaction has its own *threshold*, the lowest energy when a given process becomes energetically allowed. With the absorption of a neutron with energy in eV–keV range, the nucleus becomes excited just slightly above neutron threshold. Such, and lower, energies can be studied also with other projectiles and gamma rays.

Absorbed low-energy neutrons can be elastically reflected or continue their life inside the nucleus by first being captured to an appropriate empty nuclear orbital around threshold. The further chain of events, partly described in the previous section, can develop in different ways. The neutron interacts with nucleons inside the target; this gives rise to its loss of energy and corresponding excitation of the host nucleus. At any point, the excitation energy can be radiated out by a photon, an (n, γ) reaction, when the nucleus is left with one extra neutron in the ground or still excited state, in the last case with a perspective of the next gamma radiation. If this does not happen, after several acts of the internal nucleon–nucleon interactions, the initial excitation energy is spread over many nuclear degrees of freedom, an analog of thermal equilibration discussed earlier. But the existence of this stage does not continue forever: finally, energy can again happen to be concentrated on a single neutron that can evaporate. This elastic scattering differs from the elastic process when the neutron was not captured by the nucleus (a *direct reaction*, just an interaction with the mean nuclear potential). In the case of a relatively long lifetime of the neutron inside the nucleus, we have a quasistationary state, the subject of our previous interest (when we neglected the decay width assuming the level energy to be exactly defined). From the outside, the whole long-lived story, from neutron capture to evaporation, can be observed as a rather narrow *neutron resonance*.

The total width Γ of the resonance on the energy scale corresponds to the mean lifetime of the intermediate system with respect to all possible decay processes and, in the case of a mean lifetime τ, can be estimated as

$$\Gamma \sim \frac{\hbar}{\tau}. \tag{6.1}$$

For τ greater than the characteristic time required for several acts of intermediate interactions inside the nucleus, the equilibration process lasts long enough to acquire a complicated wave function and a system is called, after Niels Bohr, a *compound nucleus* [41], see Fig. 1.2. In this case, the return of a neutron to the continuum indeed reminds the evaporation process [249]. With a spectrum of incoming neutrons of different (still very low) energy, we go along the energy axis through a chain of quite close but non-overlapping resonances, Fig. 6.1. The energy distribution of evaporated neutrons is close to Maxwellian with the effective temperature defined by the degree of excitation of the residual nucleus [142].

With growth of the starting energy, gradually, the lifetime of the nucleus after absorption of the neutron is getting shorter, the widths of the corresponding resonances grow and they start overlapping. However,

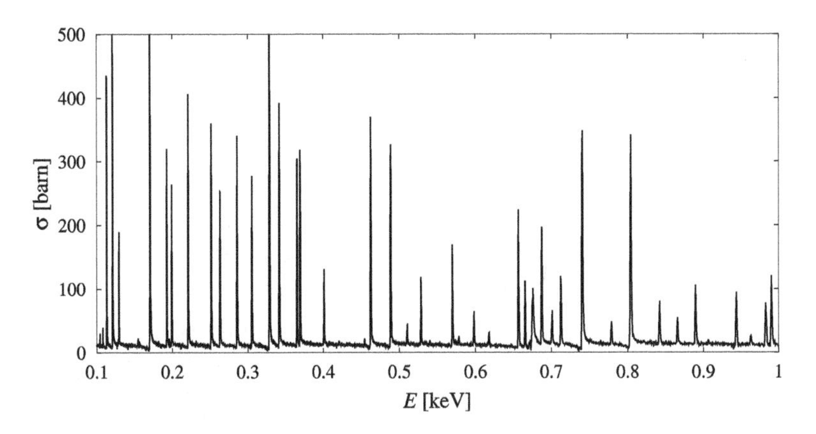

Fig. 6.1. A piece of the neutron resonance spectrum for ^{232}Th+n as a function of energy, experimental data from Ref. [98].

the sequence of resolved neutron resonances at low energy is, in many practical examples, quite long. If the resonances are not infinitely narrow, we speak about the spectrum of their centroids. Taking a resonance spectrum (or, as it was done long ago in Ref. [40], combining spectra from different nuclei), one can try to study the statistics of resonances and juxtapose them to the predictions of random matrix theory. Indeed, this comparison reveals the GOE-type statistics of level spacings and other related properties. This is literally possible only when the widths of resonances are small and they can approximately mimic stationary states. Essentially, this is what we have discussed in the previous chapter with unstable quantum states obtained in experiments and treated as if they would have been stationary. Then it was possible to compare their statistics with ideal random matrix ensembles and with the shell model.

Of course, apart from neutron scattering, there are numerous types of nuclear reactions which provide information that can be analyzed keeping in mind the mesoscopic aspects. The system still consists of many interacting quantum constituents that brings to the light various aspects of quantum chaos. But at some point the intermediate character of excited nuclear states in the reaction process becomes really important so that the application of stationary random matrix ensembles is getting not sufficient. In the processes taking place not far from thresholds, the probability of multiple excursions into continuum and back has to be accounted for. With necessity, we come to *open* mesoscopic systems.

6.2. Formal Description of an Open Quantum System

Here we introduce the theoretical approach sometimes called *Continuum Shell Model* that provides a good chance to describe in the common framework reactions and intrinsic properties of an open mesoscopic system, such as a nucleus, including the features where the influence of the continuum becomes noticeable or even crucial. This formalism is based on the *Feshbach projection method* [87] that allows one to exclude the details of the continuum wave functions from the explicit consideration of a process but fully account for their influence onto the properties of the intrinsic states which are stable or would be stable without presence of the continuum. This approach is capable also to consider reactions themselves but this is not our main task in this chapter although comments will be given.

The mathematics of the method uses the standard identity (3.13) for real x and η,

$$\lim_{\eta \to +0} \frac{1}{x \pm i\eta} = \text{P.v.}\, \frac{1}{x} \mp i\pi\delta(x). \tag{6.2}$$

The principal value (P.v.) term will describe the virtual *off-shell* continuum influence on the properties of many-body states that leads to the *Thomas–Ehrman* energy shift [80, 227] studied originally from the viewpoint of violation of isospin symmetry in nuclei [15]. The second term in (6.2) gives rise to the real *on-shell* processes of exciting intrinsic states which finally decay back into continuum, where the delta-function guarantees the energy conservation in the reaction.

Now we apply the Feshbach formalism to a nuclear reaction [154] enacted by the total (including the target, projectile, and products of the reactions) Hermitian Hamiltonian $\hat{H}$. Here we follow mainly the first discussions related to the mesoscopic aspects of the continuum problem given in [208, 209, 211]; we can also refer to the later reviews [18, 241, 245, 262]. In the stationary formulation, we start with the Schrödinger equation for the total many-body wave function with certain energy E,

$$\hat{H}\Psi_E = E\Psi_E, \tag{6.3}$$

where the complete basis in the Hilbert space contains wave functions with different asymptotics: bound, $|k\rangle$, and continuum, $|c;E\rangle$, states with their characteristic threshold energies E^c for opening the channel c. The superscript ^{c} carries all quantum numbers of the channel. In principle, the threshold energies E^c should come self-consistently from the solution of the many-body problem. Only then the genuinely bound states will be determined, while some states, preliminary considered as bound, will turn out to be unstable being populated in some reactions and then decaying back into continuum. This redevelopment occurs through the physical interaction and mixing between the two original classes of states.

Following the Feshbach formalism we divide the Hilbert space into two subspaces, $\mathcal{P}$ with asymptotics of waves in the continuum and $\mathcal{Q}$ corresponding to states bound in the absence of the interaction between the classes. The corresponding subdivision provides the Hamiltonian matrix consisting of four submatrices,

$$\hat{H} = \hat{H}_{\mathcal{Q}\mathcal{Q}} + \hat{H}_{\mathcal{Q}\mathcal{P}} + \hat{H}_{\mathcal{P}\mathcal{Q}} + \hat{H}_{\mathcal{P}\mathcal{P}}. \tag{6.4}$$

Here $\mathcal{Q}$ and $\mathcal{P}$ are projection operators for corresponding subspaces. General state vectors have components in both classes,

$$\Psi = \Psi_{\mathcal{Q}} + \Psi_{\mathcal{P}}, \tag{6.5}$$

satisfying the obvious coupled equations

$$(\hat{H}_{\mathcal{Q}\mathcal{Q}} - E)\Psi_{\mathcal{Q}} = -\hat{H}_{\mathcal{Q}\mathcal{P}}\Psi_{\mathcal{P}}, \quad (\hat{H}_{\mathcal{P}\mathcal{P}} - E)\Psi_{\mathcal{P}} = -\hat{H}_{\mathcal{P}\mathcal{Q}}\Psi_{\mathcal{Q}}. \tag{6.6}$$

Eliminating, for example, the $\mathcal{P}$ part, we come to the equation for the $\mathcal{Q}$ part,

$$\hat{\mathcal{H}}(E)\Psi_{\mathcal{Q}} = E\Psi_{\mathcal{Q}}, \tag{6.7}$$

with the effective *energy-dependent* Hamiltonian in the $\mathcal{Q}$ space

$$\hat{\mathcal{H}}(E) = \hat{H}_{\mathcal{Q}\mathcal{Q}} + \hat{H}_{\mathcal{Q}\mathcal{P}} \frac{1}{E - \hat{H}_{\mathcal{P}\mathcal{P}}} \hat{H}_{\mathcal{P}\mathcal{Q}}. \tag{6.8}$$

In the preceding paragraph, the subdivision of the space into two parts was just formal and in fact arbitrary revealing the generality of the approach as stressed by Feshbach himself [88]. Let some states of class $\mathcal{P}$ belong to the continuum part of the spectrum. If our energy E is above threshold for a certain channel of the $\mathcal{P}$ space, the denominator in Eq. (6.8) displays the singularity. The appropriate way of handling this singularity is to consider the energy E in this equation as a limit from the upper half of the complex energy plane, $E \Rightarrow E^{(+)} = E + i\eta$, $\eta \to +0$. This will lead the states of the former $\mathcal{Q}$ space existing at energy above threshold to acquire the positive width Γ describing the irreversible decay. Following [82], we assume that we use the *eigenchannel* basis in the $\mathcal{P}$ space that brings the part $\hat{H}_{\mathcal{PP}}$ to the diagonal form. Then Eq. (6.8) leads to the matrix equation in the $\mathcal{Q}$ space,

$$\hat{\mathcal{H}}_{12}(E) = H_{12} + \sum_{c} \int d\tau\, dE' \,\langle 1|\hat{H}_{\mathcal{QP}}|c, \tau, E'\rangle \times \frac{1}{E^{(+)} - E'(c, \tau)} \langle c, \tau, E'|\hat{H}_{\mathcal{PQ}}|2\rangle. \tag{6.9}$$

Here the basis states in the $\mathcal{Q}$ space are labeled by simple numbers, while the variables in channel c are hidden in the generalized symbol τ.

The singularity in channel c emerges at energy E higher than threshold E^c in this channel. This is the right time to recall the identity (6.2) and separate the principal value and the delta-function contribution in such dangerous denominators. This leads to the working form of the effective non-Hermitian and energy-dependent Hamiltonian,

$$\hat{\mathcal{H}} = \hat{\hat{H}} - \frac{i}{2}\hat{W}(E). \tag{6.10}$$

The new Hermitian part $\hat{\hat{H}}$,

$$\hat{\hat{H}} = \hat{H}_{\mathcal{QQ}} + \hat{\Delta}(E), \tag{6.11}$$

consists of the original intrinsic Hamiltonian and the energy-dependent part $\hat{\Delta}(E)$ coming from the principal value contributions of the $\mathcal{P}$ sector, the *off-shell* (virtual) couplings leading to renormalizations of internal states. Here all channels, open and closed at a given energy E, contribute, similarly to the standard perturbation theory in the discrete spectrum. The anti-Hermitian part $\hat{W}(E)$ in Eq. (6.10) comes from the delta function term in (6.2) and describes the coupling to the continuum through the transitions into channels open at this energy. Including in the matrix elements of $\hat{W}$ the factors coming from kinematic variables in a given channel, we can write this part of the Hamiltonian in a general factorized form,

$$W_{12}(E) = 2\pi \sum_{c(\text{open})} A_1^c(E) A_2^{c*}(E), \tag{6.12}$$

with the sum over channels open at energy E.

6.3. Relation to the Scattering Matrix

The amplitudes $A_1^c(E)$ form an $N \times M$ matrix $\mathbf{A}$ describing the quantum transitions between internal states $k = 1, 2, ..., N$ and those reaction channels $a, b, \ldots, M$ which are open at energy E. The part (6.12) of the effective Hamiltonian $\hat{\mathcal{H}}$ can be depicted as

$$\hat{W} = 2\pi\, \mathbf{A}\mathbf{A}^{\dagger}; \tag{6.13}$$

in a time-reversal invariant case, the amplitudes A_1^c can be taken as real, $\mathbf{A}^{\dagger} \Rightarrow \mathbf{A}^{T}$. The effective Hamiltonian taken at a complex energy $\mathcal{E}$ produces a propagator (Green's function)

$$\hat{\mathcal{G}}(\mathcal{E}) = \frac{1}{\mathcal{E} - \hat{\mathcal{H}}}. \tag{6.14}$$

Here we use the complex argument $\mathcal{E}$ while physical energy on the real axis corresponds to the limit from above in the complex plane, $\mathcal{E} \Rightarrow E^{(+)}$. The propagation inside the system without entering the channels is described by the Green's function

$$\hat{G}(\mathcal{E}) = \frac{1}{\mathcal{E} - \hat{\hat{H}}} \tag{6.15}$$

with the internal Hamiltonian (6.11).

The single propagation of the reaction signal through the system including all intrinsic interactions and corrections from the virtual coupling to the continuum is described by the $M \times M$ matrix in the channel space

$$\hat{K}(\mathcal{E}) = \mathbf{A}^{\dagger}\hat{G}(\mathcal{E})\mathbf{A}. \tag{6.16}$$

This is an analog of the R-matrix in the standard nuclear reaction theory [72, 143]. The full propagation inside the system with multiple excursions to channels and returns back is described by the series

$$\hat{L}(\mathcal{E}) = \frac{1}{1 + (i/2)\hat{K}(\mathcal{E})}. \tag{6.17}$$

Then the full propagator inside the system (6.14) is the result of summation of all those processes,

$$\hat{\mathcal{G}}(\mathcal{E}) = \hat{G}(\mathcal{E}) - \frac{i}{2}\hat{G}(\mathcal{E})\mathbf{A}\hat{L}(\mathcal{E})\mathbf{A}^{\dagger}\hat{G}(\mathcal{E}). \tag{6.18}$$

Finally, the whole history of the reaction signal propagating through the system with an arbitrary number of excursions to the continuum and back is described by

$$\hat{T}_{\mathcal{Q}}(\mathcal{E}) = \mathbf{A}^{\dagger}\hat{\mathcal{G}}(\mathcal{E})\mathbf{A} = \hat{K}(\mathcal{E})\hat{L}(\mathcal{E}) = \frac{\hat{K}(\mathcal{E})}{1+(i/2)\hat{K}(\mathcal{E})}. \tag{6.19}$$

This part of the scattering matrix,

$$\hat{S}_{\mathcal{Q}}(\mathcal{E}) = 1 - i\hat{T}_{\mathcal{Q}}(\mathcal{E}) = \frac{1-(i/2)\hat{K}(\mathcal{E})}{1+(i/2)\hat{K}(\mathcal{E})}, \tag{6.20}$$

is evidently unitary at physical energy $\mathcal{E} \Rightarrow E^{(+)}$. One can trace the origin of unitarity to the factorized form of the continuum coupling (6.13) as was shown long ago [77].

In order to establish the connection with a usual resonance description in the reaction theory, we briefly consider the projection of the dynamics onto the external, $\mathcal{P}$, space [86]. Instead of (6.8), we have then

$$\hat{\mathcal{H}}_{\mathcal{P}}(E) = \hat{H}_{\mathcal{PP}} + \hat{H}_{\mathcal{PQ}}\frac{1}{E-\hat{H}_{\mathcal{QQ}}}\hat{H}_{\mathcal{QP}} \tag{6.21}$$

with energy E in the continuous spectrum and therefore with the presence of solutions $\psi^{c}_{E'}$ for the homogeneous equation $(H_{\mathcal{PP}} - E')\Psi_{\mathcal{P}} = 0$. To illustrate the situation, take just one intrinsic state Φ with unperturbed energy $H_{\mathcal{QQ}} \Rightarrow E_0$ coupled to a single channel. The general solution at energy E will be a superposition

$$\Psi_E = a(E)\Phi + \int dE'\, b(E,E')\psi_{E'}. \tag{6.22}$$

The continuum coupling amplitude is, in our notations,

$$A(E') = \langle\Phi|\hat{H}_{\mathcal{QP}}|\psi_{E'}\rangle. \tag{6.23}$$

The amplitudes in Eq. (6.22) satisfy the set of coupled equations

$$(E-E_0)a(E) = \int dE'\, A(E')b(E,E'), \quad (E-E')b(E,E') = A^{*}(E')a(E). \tag{6.24}$$

In the presence of the homogeneous solution at $E' = E$, the amplitudes $b(E, E')$ have a general form (6.2) of the delta function with some energy-dependent real amplitude $r(E)$ and the principal value part,

$$b(E, E') = \left[\text{P.v.}\, \frac{1}{E - E'} + r(E)\delta(E - E')\right] A^*(E')a(E). \tag{6.25}$$

From (6.24), accounting for the principal value integral that, in the general case of energy-dependent coupling amplitudes $A(E)$, gives rise to the shift $\Delta(E)$, we find

$$r(E) = \frac{E - E_0 - \Delta(E)}{|A(E)|^2}. \tag{6.26}$$

As shown by Fano [86], this determines the phase shift in the reaction, $\tan\delta(E) = -\pi/r(E)$. The remaining amplitude $a(E)$ can be found from the normalization in the continuum; for example, the normalization to the delta function of energies leads to

$$|a(E)|^2 = \frac{|A(E)|^2}{[E - E_0 - \Delta(E)]^2 + \Gamma^2/4}. \tag{6.27}$$

The solution has a resonance shape with the energy position $E_0 + \Delta(E)$ and the width,

$$\Gamma(E) = 2\pi|A(E)|^2, \tag{6.28}$$

both depending on energy of the experiment. The centroid position is determined by the maximum of the coupling amplitude. This consideration predicts asymmetric shapes of the resonances as functions of energy; in [86] this was illustrated by the autoionization of the helium atom. This asymmetry is known in nuclear near-threshold reactions as well [263].

6.4. Road to Superradiance

The effect of superradiance from an atomic system was predicted by Dicke [75] and for the first time observed in an optically pumped HF gas [206]. Later it was studied in detail and, already as a more general physical phenomenon, widely used in quantum optics, atomic, plasma, and beam physics, see, for example, [8, 34, 130]. There are also interesting applications [63] to the systems with coexistence of superconductivity and superradiance.

A simplified model of the phenomenon [18] can be presented by an image of a cavity filled with identical two-level "atoms". An excited atom can irreversibly radiate from the cavity making the transition between atomic levels. Along with that, a radiated quantum can be absorbed in resonance by an identical atom inside the cavity. Such processes of absorption and reemission create a special *superradiating* collective state as a coherent combination of $N \gg 1$ atoms coupled by the resonance signal. In this state, the contribution of each individual atom is of the order $1/\sqrt{N}$ but the coherence of these amplitudes gives a factor N that corresponds to the burst of radiation. If we start with all N atoms in the excited state, the ongoing deexcitation process may lead to the state where occupancy amplitudes of the ground and excited atoms are equal and this is the best condition for establishing the coherent coupling due to the enhanced oscillator amplitudes for both induced radiation and absorption.

The onset of the collective radiation is formally the consequence of the general factorized structure (6.12) of the effective Hamiltonian in the intrinsic space. Let us use the eigenbasis $|\alpha\rangle$ of interacting intrinsic eigenstates of the real part of the effective Hamiltonian (6.10),

$$\hat{\tilde{H}}|\alpha\rangle = \epsilon_\alpha|\alpha\rangle. \tag{6.29}$$

In this *internal representation*, the imaginary part of the total Hamiltonian is still factorized as in Eq. (6.12), so that the matrix of the total effective Hamiltonian is

$$\mathcal{H}_{\alpha\beta} = \epsilon_\alpha\delta_{\alpha\beta} - \frac{i}{2}2\pi \sum_{c(\text{open})} A^c_\alpha A^{c*}_\beta. \tag{6.30}$$

The eigenstates of $\mathcal{H}$ are in general quasistationary states with complex energies $\mathcal{E} = E - (i/2)\Gamma$ above reaction thresholds and stationary states with real energies ($\Gamma = 0$) below thresholds.

For the factorized (even if imaginary) interaction term in the Hamiltonian, the formal situation is quite similar to what was discussed in relation to a Hermitian collective excitation, Sec. 2.8. There are characteristic limits for the behavior of the system. The main parameter is the ratio of the typical level width Γ to the level spacing D on the real energy axis, $\kappa = \Gamma/D$, or the ratio of traces of imaginary and real parts of the Hamiltonian $\mathcal{H}$. Here we assume that this ratio indeed describes the average behavior of the system as many states have close values of this parameter. At weak coupling to continuum, $\kappa \ll 1$, we deal with rather narrow separated resonances. This situation occurs for neutron resonances at very low energy when the nuclear state after the capture of a slow neutron proceeds through a protracted stage of thermalization. Then the lifetime is long, the imaginary corrections to real resonance energies are small, the quasistationary states still can be labeled by their internal names $|\alpha\rangle$, and the width can be evaluated by the diagonal element of W, Eq. (6.10),

$$\mathcal{E}_r \Rightarrow \mathcal{E}_\alpha = \epsilon_\alpha - \frac{i}{2}\Gamma_\alpha, \tag{6.31}$$

where

$$\Gamma_\alpha = \sum_c \gamma^c_\alpha, \quad \gamma^c_\alpha = 2\pi |A^c_\alpha|^2. \tag{6.32}$$

The actual distribution of widths for a certain channel c requires separate consideration and we will return to this question later.

With increasing continuum coupling, the resonances start to overlap, and the parameter κ becomes greater than 1. The complex energies of quasistationary states (in this limit they are broad resonances), according to Eqs. (6.18)–(6.20), are the roots of

$$\text{Det}\left\{1 + \frac{i}{2}\hat{K}(\mathcal{E})\right\} = 0. \tag{6.33}$$

For example, for one open channel, the resonances are the complex roots of

$$f(\mathcal{E}) = 1 + \frac{i}{2}\sum_\alpha \frac{\gamma_\alpha}{\mathcal{E} - \epsilon_\alpha} = 0. \tag{6.34}$$

The previous limit of independent narrow resonances gives here the same result (6.31), $\Gamma_\alpha = \gamma_\alpha$, the diagonal elements of the matrix W.

This situation is formally similar to the collectivization of states due to the real factorized interaction that is known to lead to the appearance of a giant resonance accumulating the significant part of the collective strength of given symmetry, Sec. 2.8. But here, the collectivization takes place along the imaginary axis in the complex energy plane and brings in the state of a collectivized decay, an analog to superradiance. In the extreme limit, when the external mixing is very strong and the widths noticeably overlap neighboring intrinsic levels, the part W of the effective Hamiltonian (6.10) plays the main role. Being factorized, the operator W has in the one-channel case just one non-zero eigenvalue equal to the trace,

$$w = \text{Tr}\, W \sim N\gamma. \tag{6.35}$$

This is the superradiant limit,

$$\mathcal{E} \approx \bar{\epsilon} - \frac{1}{2} w, \tag{6.36}$$

where $\bar{\epsilon}$ is a centroid of the levels along the real energy axis. The remaining $N - 1$ states are *trapped* with their widths going to zero (the trace w saturates the whole imaginary part of the total Hamiltonian).

More frequently, we have to deal with intermediate situations that can be realized in actual nuclear physics related to giant resonances [151, 210]. If the continuum interaction is strong enough, it might be convenient to consider the problem of the total Hamiltonian $\hat{\mathcal{H}}$ in the basis of its anti-Hermitian part W. In the matrix for one-channel case, we make the first the state that concentrates the total imaginary part $-(i/2)w$. The remaining submatrix of dimension $N - 1$ is Hermitian. It can be diagonalized to get its real eigenvalues $\tilde{\epsilon}_\nu$, where $\nu = 2, ..., N$. The element $\mathcal{H}_{11}$ has some real part h, while the off-diagonal elements $H_{1\nu} \equiv h_\nu$ in this representation occupy only the upper row and the left column of the matrix. In fact, the diagonal elements $\tilde{\epsilon}_\nu$ are interlocated with ϵ_α of the previous basis. The secular equation for complex energies $\mathcal{E}$ now is

$$\mathcal{E} - h - \mathcal{R}(\mathcal{E}) + \frac{i}{2} w = 0, \quad \mathcal{R}(\mathcal{E}) = \sum_{\nu=2}^{N} \frac{|h_\nu|^2}{\mathcal{E} - \tilde{\epsilon}_\nu}. \tag{6.37}$$

If the continuum mixing is strong, we are on the way to the previous case, and coupling amplitudes h_ν are small. The superradiating state is not an exact solution anymore but it serves as a *doorway state* bringing a small decay chance described by the function $\mathcal{R}$ to the sea of previously stable states $\nu \geq 2$. If h_ν can be considered a perturbation, in the lowest order we find

$$\mathcal{E}_1 = h - \frac{i}{2} w, \tag{6.38}$$

$$\mathcal{E}_\nu = \tilde{\epsilon}_\nu + \frac{h_\nu^2}{\tilde{\epsilon}_\nu - h + (i/2)w}, \quad \nu \geq 2. \tag{6.39}$$

The widths corresponding to those complex energies are

$$\Gamma_1 = w \left[1 - \sum_{\nu \geq 2} \frac{h_\nu^2}{(h - \tilde{\epsilon}_\nu)^2 + (1/4)w^2} \right], \tag{6.40}$$

$$\Gamma_\nu = w \frac{h_\nu^2}{(h - \tilde{\epsilon}_\nu)^2 + (1/4)w^2}, \quad \nu \geq 2. \tag{6.41}$$

In this limit, the parameter $\kappa \gg 1$, so that the bright (Dicke doorway) state has a width of the order $w(1-\kappa^{-2})$, while the dark (trapped) states share the remaining part $\sim w/\kappa^2$ borrowed from the Dicke state, $\Gamma_\nu \propto w/(\kappa^2 N)$. The width of the doorway state almost exhausts the total width available. With the mean distance between the levels (resonance centroids on the real axis) $D \approx \Delta E/N$, the ratio $\Gamma_\nu/D \sim 1/\kappa \ll 1$, and the narrow long-lived resonances do not overlap. It is important that the picture of many narrow resonances on the background of the broad state appears naturally due to the strong coupling to the continuum being in fact dictated by the unitarity of the matrix W.

In a more general case of k open channels with their continuum couplings of the same order of magnitude, there will be formed k broad resonances practically absorbing the total width. This trend was noticed long ago (but not immediately understood) in the realistic calculations [134] of resonance processes in nuclei with strong continuum coupling. At $N \gg 1$, the transition from the picture of similar isolated resonances to broad coherent states is rather sharp in the vicinity of $w/\Delta E \geq 1$ looking as a phase transition.

There is a large-scale hierarchical sequence of time intervals appearing in the nuclear reaction physics. The characteristic time for formation of the compound nucleus after the initial excitation in a reaction is the *Weisskopf time* $\sim \hbar/D$ necessary to resolve the fine energy structure of the spectrum. The shorter time for fragmentation of the initial doorway state into more complex configurations of a given type can be estimated as $\hbar/\Delta E$ where ΔE is determined by the number of principal components of the collective state considered in the exact stationary basis of the internal Hamiltonian. The even shorter time $\hbar/w$ corresponds essentially to a direct process. With notations used in nuclear reaction theory, this means that $\Gamma^{\uparrow} \sim w \gg \Gamma^{\downarrow} \sim \Delta E$. As seen from (6.41), at $w \gg \Delta E$, the metastable states have a typical lifetime $\tau \sim (w/\Delta E)\hbar/D$ that is longer than $\hbar/D$ so that there is enough time for establishing thermal equilibrium in a compound nucleus. As an example, one can consider elastic scattering with formation of a single-particle resonance coupled with the sea of stable many-body states [231].

The mechanism of appearance of superradiating states in a many-body system can be illustrated by a simple model, Fig. 6.2. Let us put N fermions in the Hilbert space of Ω single-particle levels, where the upper level is already in the continuum. Due to the Pauli exclusion principle, there are $\Omega!/[N!(\Omega-N)!]$ allowed many-body configurations. These configurations acquire the continuum width if the quasistationary upper level is partially occupied in a given configuration. When the decay probability of the upper

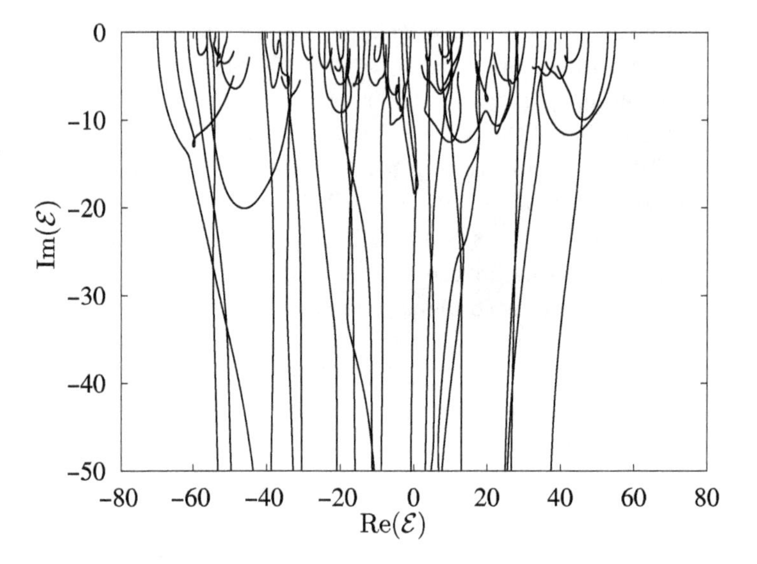

Fig. 6.2. Complex plane trajectories of energies and widths of many-body states in a system of three fermions on eight levels as the decay amplitude from the upper level is increased.

level is artificially increased, all such configurations get noticeable widths and their energies become complex, Fig. 6.2. However, after some increased value of the continuum coupling, many level trajectories in this complex energy plane return to the real axis while few of them continue to fall deeply parallel to the imaginary axis describing the growing widths. Those are superradiating states. They correspond to the many-body configurations where the decaying single-particle level is occupied. We can predict the number of such decaying trajectories: it is equal to the number of the many-body states where $N-1$ particles occupy remaining $\Omega-1$ levels while the level in the continuum is occupied with probability one. This schematic example already shows that the continuum coupling may become the main driving force of the many-body dynamics.

In a system of many intrinsic states coupled to many decay channels, the appearance of the superradiating states with the increase of the continuum coupling looks as a genuine sharp phase transition [147]. Figure 6.3 shows the modeling of a system with $N=4000$ states coupled to 1000 decay channels. The Hermitian part of the Hamiltonian is modeled by a realization from a Gaussian Orthogonal Ensemble with the parameter $\lambda=1$, while the non-Hermitian part $\hat{W}=\varkappa^2\mathbf{A}\mathbf{A}^\dagger$ is modeled by 1000 orthonormal vectors A^c. The parameter $\varkappa$ is varied corresponding to different strengths of

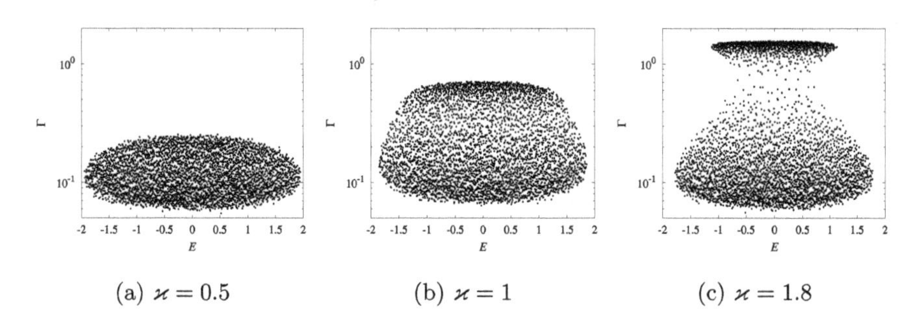

(a) $\varkappa = 0.5$ (b) $\varkappa = 1$ (c) $\varkappa = 1.8$

Fig. 6.3. Emergence of a cloud of super-radiant states with the growth of continuum coupling.

the imaginary part. At relatively weak continuum coupling, $\kappa = 0.2$, all widths (imaginary parts of the energy eigenvalues) form a dense cloud in the lower part of the complex energy plane with a modest spread inside this cloud. With κ growing, the fast decaying part of the cloud starts separating from the main complex. At a large value of κ, this part, containing exactly a quarter of all points, fully separates forming the community of superradiating states.

6.5. Collective Mode in the Continuum

This problem was already mentioned in the previous text. Here we will look in more detail at the fate of a collective mode that, due to the sign of the involved interaction, is repelled up from the corresponding shell-model energy and turned out to be located above the continuum threshold.

The first example is certainly the discussed earlier *giant dipole resonance* (GDR) that is, roughly speaking, a collective motion of neutrons relative to protons that, in the simplest realization, has a dipole character (other multipolarities are also possible but their degrees of collectivity are lower). Here we need to recall that the common (isoscalar) motion of neutrons and protons would be simply a center-of-mass displacement rather than an intrinsic nuclear excitation; in the shell-model level density problem we had to apply special efforts in order to exclude such a mode from the list of internal quantum states (we mentioned that this is necessary for the correct account for intrinsic level density [124]). As was mentioned earlier, this collective mode is not sensitive to the details of the structure so that it can be excited on top of any other low-lying nuclear state (Brink–Axel hypotheses [22]). Another ubiquitous type of nuclear excitation is the *isobaric analog state* (IAS) that can be produced, for example, in the case of the Fermi-mode of weak decay, simply by the isospin operators $T_{\pm}$, but in the neighboring nucleus that belongs to the same isobaric multiplet.

It is important to understand the difference between the original derivation of the collective wave function as a coherent superposition of simple mean-field (particle-hole) excitations and the subsequent expansion of this combination over the *exact* stationary states which are, in this energy region, chaotic eigenstates of the full many-body Hamiltonian. This is a well-known manifestation of the concept of *quasiparticles* (in this case phonons) with the finite lifetime [173]. Here we are interested in the situation when the underlying components (at least part of them) are already in the continuum. When an experiment provides the curve of excitation of the resonance collective state as a function of energy, the width Γ of this curve, related to the mean lifetime by the uncertainty relation, reflects both mechanisms, internal damping, $\Gamma^{\downarrow}$, and irreversible decay into continuum, $\Gamma^{\uparrow}$. In all cases, the manifestations of the simple mode in specific reaction channels are intertangled with the chaotic mixing inside the system.

Here it is appropriate to use the effective non-Hermitian Hamiltonian (6.10) in order to take into account internal and external interactions on equal footing. The internal structure at high level density produces the *background* of the intrinsic basis states $|n\rangle$, where $n = 1, \dots, N$ and N is supposed to be large; their energies with no coupling to the collective state are h_n. The original simple state $|0\rangle$ with unperturbed energy ϵ_0 is typically located in the same range of energy. All $(N+1) \gg 1$ states have the same values of exact characteristic constants of motion, such as the total angular momentum. The anti-Hermitian part W has a special structure (6.13) being originated by the on-shell decays into open channels $c = 1, \dots, M$. The Hermitian part $\hat{H}$ consists of the unperturbed energy ϵ_0 of the simple state $|0\rangle$, the internal $N \times N$ Hamiltonian $\hat{h}$ describing the background states $|n\rangle$, and the intrinsic coupling $H_{0n} \equiv V_n$ between the simple and complicated states.

It is useful first to recall the solution of the problem *without continuum*. It is given by the diagonalization of the internal Hamiltonian, when its eigenvalues are the roots $E = \epsilon_\alpha$ of the secular equation

$$X(E) \equiv E - \epsilon_0 - \sum_{n=1}^{N} \frac{|V_n|^2}{E - h_n} = 0. \tag{6.42}$$

The wave functions $|\alpha\rangle$ of the internal part are the superpositions

$$|\alpha\rangle = C_0^\alpha|0\rangle + \sum_{n=1}^{N} C_n^\alpha|n\rangle, \quad |C_0^\alpha|^2 + \sum_{n=1}^{N}|C_n^\alpha|^2 = 1, \tag{6.43}$$

where the weight of the collective state is

$$f^\alpha = |C_0^\alpha|^2 = \left(\frac{dX}{dE}\right)^{-1}_{E=\epsilon_\alpha} = \left[1 + \sum_{n\geq 1}\frac{|V_n|^2}{(\epsilon_\alpha - h_n)^2}\right]^{-1}. \tag{6.44}$$

With the normalization $\sum_\alpha f^\alpha = 1$, this describes the spreading of the collective state over exact stationary states of the intrinsic dynamics. On the other hand, this shows the presence of a *scar* [131] as a memory of the original collective mode infecting the same symmetry states in this energy region. In many realistic applications [38, 145, 148], the result of this interaction can be in average described by the *uniform model* leading to the Breit–Wigner *spreading width* that is what we have called $\Gamma^\downarrow$ (assumed to be greater than D),

$$f(\epsilon) = \frac{1}{2\pi}\frac{\Gamma^\downarrow}{(\epsilon - \epsilon_0)^2 + (\Gamma^\downarrow)^2}. \tag{6.45}$$

The spreading width is given by the *golden rule* with the mean value of the coupling amplitude squared,

$$\Gamma^\downarrow = 2\pi\frac{\langle V^2\rangle}{D}. \tag{6.46}$$

Such a description is valid if the spreading width does not exceed the energy range ΔE of coupling strength V_n^2 defined by the spread of the doorway states which provide the gates for the further mixing of the original state $|0\rangle$. This is expected to be a good approximation for the IAS with the typical spreading width $\leq$ 100 keV. In the case of giant resonances, the model should be corrected [145] but the difference influences mainly the wings of the strength function which are of minor importance for our purpose; here we use the uniform model for definiteness.

Now we take into account the openness of the system [213]. The collective state $|0\rangle$ is open to direct decay channels c displaying specific signatures of the simple mode, for example, collective gamma radiation of a certain multipolarity from the giant resonance, or pure isospin of the IAS.

Because of the intrinsic coupling to compound states, the collective states also acquire access to many *evaporation channels* labeled below by the subscript "ev"; partial widths depend on the strength distribution of the simple mode over specific compound states.

To describe the open compound states, we introduce the $N \times N$ Green's function

$$\hat{g}(z) = \frac{1}{z - \hat{h} + (i/2)\hat{w}}, \tag{6.47}$$

where $\hat{w}$ stands for the $N \times N$ submatrix of W that acts in the compound subspace and describes the evaporation together with the interaction between the compound states through common decay channels. The latter is characterized by the off-diagonal matrix elements of $\hat{w}$. N complex poles z_n of $g(z)$ determine energies and evaporation widths of compound resonances still decoupled from the simple mode. Being the sums of uncorrelated contributions of many evaporation channels, $M \gg 1$, these elements, due to mutual cancelations, are typically small by a factor $1/\sqrt{M}$ in comparison with the diagonal elements [213]. The corresponding corrections are of order of $\gamma_{\text{ev}}^2/(MD^2)$ where γ_{ev} is the typical evaporation width. We will neglect them below assuming $\gamma_{\text{ev}} \ll \sqrt{M}D$. Under this condition, partial decay widths of the compound states to specific evaporation channels are small, $\gamma_{\text{ev}}/M \ll D$. With those assumptions, the complex energies of compound resonances, $\nu = 1, \ldots, N$, are equal to $\tilde{\epsilon}_\nu = h_\nu - (i/2)\gamma_{\text{ev}}$, $n = 1, 2, \ldots, N$, assuming on statistical grounds that the width fluctuations of compound states are weak if the number M of open evaporation channels is large. The original simple (collective) state, before being coupled to compound states, had its own complex energy $\tilde{\epsilon}_0 = \epsilon_0 - (i/2)\gamma_0$, where γ_0 is the direct decay width.

Now we switch on a friction-type interaction between the simple and compound states through the Hermitian coupling operator $\hat{V}$. The mixing proceeds in competition with decays, both via direct and evaporation channels. The diagonalization of the total non-Hermitian Hamiltonian leads to $N+1$ complex eigenvalues which are the roots $z = \mathcal{E}_j$ of the secular equation [cf. Eq. (6.42)]

$$z - z_0 - \mathbf{V}^T g(z)\mathbf{V} = 0 \quad \Rightarrow \quad \mathcal{E}_j - z_0 - \sum_\nu \frac{V_\nu^2}{\mathcal{E}_j - z_\nu} = 0. \tag{6.48}$$

The interaction amplitudes V_ν, which couple the unstable simple state with complicated (and decaying as well) intrinsic states $|\nu\rangle$, are still real in the approximation taken above as we neglected the off-diagonal part of the continuum coupling w.

Similar to Eq. (6.43), the quasistationary eigenstates $|j\rangle$ of the full problem can be presented as superpositions of decoupled unstable states $|0\rangle$ and $|\nu\rangle$,

$$|j\rangle = \tilde{C}_0^j|0\rangle + \sum_\nu \tilde{C}_\nu^j|\nu\rangle. \tag{6.49}$$

The fraction $\tilde{f}_j = |\tilde{C}_0^j|^2$ of the strength of the simple state carried by the quasistationary state $|j\rangle$ is equal to

$$\tilde{f}^j = \frac{1}{1+L^j}, \quad L^j = \mathbf{V}^T g^\dagger(\mathcal{E}_j) g(\mathcal{E}_j)\mathbf{V}. \tag{6.50}$$

With $\mathcal{E}_j = E_j - (i/2)\Gamma_j$, the loops L^j can be written as

$$L^j = \sum_\nu \frac{V_\nu^2}{|\mathcal{E}_j - z_\nu|^2} = \frac{2}{\Gamma_j - \gamma_{\rm ev}} \,\mathrm{Im} \sum_\nu \frac{V_\nu^2}{\mathcal{E}_j - z_\nu}. \tag{6.51}$$

Using the secular equation (6.48) we arrive at a very simple expression,

$$L^j = \frac{\gamma_0 - \Gamma_j}{\Gamma_j - \gamma_{\rm ev}}, \tag{6.52}$$

leading to the individual strengths (6.50)

$$\tilde{f}^j = \frac{\Gamma_j - \gamma_{\rm ev}}{\gamma_0 - \gamma_{\rm ev}}. \tag{6.53}$$

In other words, the resulting width of the quasistationary state $|j\rangle$ can be found from simple probabilistic arguments,

$$\Gamma_j = \gamma_0 \tilde{f}^j + \gamma_{\rm ev}(1 - \tilde{f}^j). \tag{6.54}$$

The direct decay width is distributed over all quasistationary states according to their fractions of the strength of the original collective state. It is easy to check the normalization of the weights:

$$\sum_j \tilde{f}^j = \frac{\sum_j \Gamma_j - (N+1)\gamma_{\rm ev}}{\gamma_0 - \gamma_{\rm ev}} = 1, \tag{6.55}$$

where the last step follows from the invariance of the imaginary part of the trace of the Hamiltonian,

$$\sum_j \Gamma_j = \gamma_0 + N\gamma_{\rm ev}. \tag{6.56}$$

The probabilistic interpretation emerged here as a result of a strict quantummechanical calculation, with no transition to a kinetic description after we have neglected the fluctuations of evaporation widths.

The explicit analytical solution for the problem of the collective state in the continuum and its coupling to chaotic states close in energy can be found in the uniform model [213] where the spectrum of underlying complicated states is taken as equidistant with the energy spacing D, evaporation width γ, and the constant squared matrix element $\langle V^2\rangle$. Then the typical sum needed in specific calculations is that for the Green's function,

$$S(E;D,\gamma) = \sum_{\nu=-\infty}^{\infty} \frac{1}{E - \nu D + (i/2)\gamma}. \tag{6.57}$$

The exact summation gives

$$S(E;D,\gamma) = \frac{\pi}{D}\frac{x - iy}{1 + ixy}, \tag{6.58}$$

where the parameters are introduced

$$x = \cot\left(\frac{\pi E}{D}\right), \quad y = \tanh\left(\frac{\pi\gamma}{2D}\right). \tag{6.59}$$

As the decay width γ increases, y rapidly changes from a small value $y \approx (\pi\gamma/2D)$ for isolated long-lived states, when $(\gamma/D) \ll 1$, to a value exponentially close to 1 for overlapping levels, when $(\gamma/D) \gg 1$.

It is often sufficient to consider just these two limiting cases. At small γ, the imaginary part of S is small, being proportional to $y \approx (\pi\gamma/2D)$, and the real part of S is equal to $\pi x/D$ as for stable levels. In the opposite case of large γ/D, the real part vanishes as $1-y^2$, whereas $\operatorname{Im} S = -\pi/D$. Both cases have a general meaning being not limited by restrictions of the uniform model. Thus, the result for the overlapping case follows immediately after substituting $\sum(E-\epsilon_\nu)^{-1}$ by the integral with a level density $1/D$ and using a small shift of energies into the complex plane. This expression is routinely used in statistical theory of nuclear reactions [121]. A similar consideration is valid for more complex sums [213].

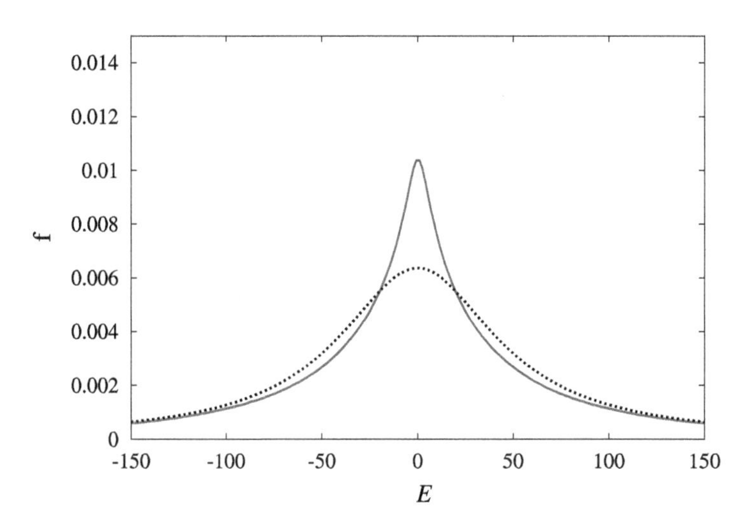

Fig. 6.4. Strength function as a function of energy, dotted line shows the Breit–Wigner approximation, see [213].

Figure 6.4 illustrates the solution of equations for the strength function $\tilde{f}^j \gg D$. The corresponding analytic expression and discussion of typical cases, from a *broad pole* in the case of the well-pronounced collective peak to the full dissolution of the collective mode in the ocean of evaporation states, can be found in [213].

6.6. Collective Strength in Reactions

Up to now we concentrated on the "inside" view of a collective mode mixed with complicated fine structure states. The "outside" world was present as a reservoir for irreversible decay of all those intrinsic states through numerous open channels. Now we take a glimpse of the same system from the viewpoint of reaction amplitudes and cross sections where only asymptotic states are observed. Certainly, this is the only possible way to obtain physical information on what is going on inside the system. Here we treat gamma radiation after the nuclear excitation as one of the reaction channels.

Following [154], we can write down the scattering matrix in the space of channels, $\hat{S}(E) = \{S^{cc'}\}$, as

$$\hat{S}(E) = \hat{s}^{1/2}[1 - i\hat{T}(E)]\hat{s}^{1/2}, \quad \hat{T}(E) = \mathbf{A}^T \mathcal{G}(E)\mathbf{A}. \tag{6.60}$$

Here $\hat{s}$ includes the potential scattering as well as channel coupling and direct reactions in the continuum. Those effects being unrelated to intrinsic dynamics are irrelevant for our purpose and $\hat{s}(E)$ can be considered as a diagonal matrix smoothly depending on energy E with the phase shift elements $\exp(2i\delta^c)$ for different channels. The Green's function

$$\mathcal{G}(z) = \frac{1}{z - \mathcal{H}}, \tag{6.61}$$

as well as the scattering matrix, has poles in the complex plane of the energy variable. It describes the propagation governed by the total Hamiltonian $\hat{\mathcal{H}}$, Eq. (6.8), and differs from the intrinsic Green's function by the anti-Hermitian part of the effective Hamiltonian. Due to the factorized structure of the operator $\hat{W}$, Eq. (6.12), it is easy to algebraically relate $\mathcal{G}(z)$ to the previous Green's function $G(z)$, as it was done in Eq. (6.18).

Many concepts of the scattering theory can be expressed in terms of various Green's functions introduced above. The general scattering wave function $|\Psi^c_E\rangle$ with the incident wave in the channel c at energy E can be presented by the superposition of intrinsic $|n\rangle$ and continuum channel $|c'; E\rangle$ components:

$$|\Psi^c_E\rangle = \sum_n a^c_n(E)|n\rangle + \sum_{c'} \int_{E^{c'}}^{\infty} dE'\, b^{cc'}(E,E')|c';E'\rangle. \tag{6.62}$$

Here E^c is the threshold energy in the channel c, while the decay amplitudes A^c_n are the matrix elements of the total original Hamiltonian between the states $|n\rangle$ and $|c; E\rangle$. As follows from the scattering theory [154], the $N \times M$ matrix $\mathbf{a}$ of the coefficients a^c_n is

$$\mathbf{a}(E) = \mathcal{G}(E)\mathbf{A}\,\hat{s}^{1/2}. \tag{6.63}$$

The diagonal elements of the $M \times M$ matrix $\mathbf{a}^\dagger(E)\mathbf{a}(E)$ determine the norm of the internal part of the wave function initiated in the channel c at energy E. This matrix characterizes the fraction of delay time in this reaction due to intrinsic resonances defined [207] in terms of the matrix

$$\hat{\tau}(E) = -i\hat{S}^\dagger(E)\,\frac{d\hat{S}(E)}{dE}. \tag{6.64}$$

The main resonance energy dependence comes through the retardation due to the internal propagator $\hat{K}(E)$, Eq. (6.16), so that

$$\hat{\tau}_{\text{res}}(E) = -\hat{s}^{-1/2}\,\frac{1}{1-(i/2)\hat{K}(E)}\,\frac{d\hat{K}(E)}{dE}\,\frac{1}{1+(i/2)\hat{K}(E)}\,\hat{s}^{1/2}. \tag{6.65}$$

In the same resonance approximation,

$$\left(\frac{d\hat{K}}{dE}\right)_{\rm res} = -\mathbf{A}^T\, G^2(E)\, \mathbf{A}. \tag{6.66}$$

Using the relation between the Green's functions G and $\mathcal{G}$, we obtain the physically clear result expressing the delay time as a consequence of the internal dynamics, Eq. (6.63), in the compound nucleus,

$$\hat{\tau}_{\rm res} = \hat{s}^{-1/2}\mathbf{A}^T\mathcal{G}^\dagger(E)\mathcal{G}(E)\mathbf{A}\hat{s}^{1/2} = \mathbf{a}^\dagger\mathbf{a}. \tag{6.67}$$

The total Green's function $\mathcal{G}$ describes the propagation in the open system and, therefore, the delay time as well. The full scattering matrix is unitary provided the potential scattering matrix $\hat{s}$ is unitary. It follows from the fact that the decay amplitudes $\mathbf{A}$ in the entrance and exit channels are the same that appear in all intermediate processes described by the total propagator $\mathcal{G}(E)$ with the aid of the effective Hamiltonian.

The simplest situation corresponds to the stable background states with no (or very weak) direct access to open channels, $\gamma_{\rm ev} \Rightarrow 0$, when the background states are involved by the internal coupling only at the intermediate stages of the reaction. Calculating the diagonal element of the resonance time delay matrix (6.67) we obtain for the normalized probability p^c_n, $\sum_n p^c_n = 1$,

$$p^c_n(E) = \frac{1}{(\tau_{\rm res}(E))^{cc}}\, |a^c_n(E)|^2. \tag{6.68}$$

The probability $p^c_0(E)$ characterizes the weight of the simple state $|0\rangle$ in the channel c. In the problem of the IAS, this quantity measures the isospin purity in a given channel. In the case of the stable background, we find

$$p^c_0(E) \equiv f(E) = \left[1 + \sum_n \frac{V^2_n}{(E-h_n)^2}\right]^{-1} = \left(\frac{dF}{dE}\right)^{-1}, \tag{6.69}$$

nothing but the continuous generalization of the strength function defined above in discrete points of the intrinsic energy spectrum. If several direct decay channels c are open, the energy behavior is identical for all of them being determined by intrinsic dynamics only.

Fig. 6.5. Illustration showing Borromean rings.

6.7. Continuum Shell Model: A Simple Example

It is clear that actual problems of nuclear structure and reactions, even in the realm of relatively low excitation energy, require the self-consistent consideration of the whole Hilbert space (bound states and continuum). In the above subsections, we demonstrated one of the possible logical quantum-mechanical approaches to the many-body physics of realistic systems with both bound and continuum states. Prior to showing a fully realistic application of the continuum shell model, we give a simple analytically treated example [240] that elucidates important aspects of this approach. It is in fact also quite practical for some situations in nuclear and molecular physics.

The so-called Borromean cases, such as ^{6}He, ^{9}Be, and ^{11}Li, refer to nuclear systems that can be considered to be made of three clusters with all two-body subsystems being unbound. (The name refers to the Borromea family of the Italian nobility that has a coat of arms, Fig. 6.6, made of three rings connected in such a way that cutting one of them immediately disconnects two remaining rings; a similar triangular symbol was known even earlier in Ireland.) Naturally, such systems are extremely sensitive to the continuum physics.

In general, one of the main trends in nuclear physics leads to an area where the conventional division into "structure" and "reactions" becomes insufficient: the two views of the process, from the inside (structure, properties of bound states, and transitions between them) and from the outside (reaction cross sections), should be recombined. Currently there are various theoretical approaches developed in the direction of this synthetic description. One of the most popular methods using the so-called "Gamow Shell Model" is based on the formal development by Berggren [37], practically started in [35, 36], and continued by many researchers. This popular approach is explained in detail along with practical examples and corresponding numerical recipes in the book [156]. Here we illustrate the underlying physics using the approach based on the introduced above Feshbach projection method and called simply the "Continuum Shell Model".

Important features of this theory can be seen already from the simple example of two internal mean-field states coupled, apart from the usual shell-model interaction, also by the interaction through a common decay channel [240]. We consider two single-particle levels immersed into continuum; their unperturbed energies are positive if the continuum threshold is put at zero energy. This is a prototype of the three-body Borromean model for ^{11}Li with the inert core of ^{9}Li and the intermediate particle-unstable isotope ^{10}Li. Each of the two active shell-model orbitals $2s_{1/2}$ and $1p_{1/2}$ can accommodate two halo neutrons with pair energies $\epsilon_{1,2}$. The pair states are quasistationary, and their decay amplitudes $A_{1,2}$ characterize the only open channel with the remaining core nucleus in the ground state plus the neutron pair in the state $J^{\Pi} = 0^{+}$ in the continuum. The decay amplitudes are energy dependent going to zero at threshold energy $E = 0$. Some aspects of this situation were discussed in [49, 212]; the full description was given in [240].

The general effective 2×2 Hamiltonian describing this two-state system is given by

$$\mathcal{H} = \begin{pmatrix} \epsilon_1 - (i/2)A_1^2 & V - (i/2)A_1 A_2 \\ V - (i/2)A_1 A_2 & \epsilon_2 - (i/2)A_2^2 \end{pmatrix} \tag{6.70}$$

Here the coupling V and decay amplitudes $A_{1,2}$ can be taken as real numbers. A straightforward diagonalization provides complex eigenvalues

$$\mathcal{E}_{\pm} = \bar{\mathcal{E}} \pm \frac{1}{2}\sqrt{X}, \tag{6.71}$$

where, using $A_{1,2}^2 = \gamma_{1,2}$,

$$\bar{\mathcal{E}} = \frac{1}{2}[\epsilon_1 + \epsilon_2 - (i/2)(\gamma_1 + \gamma_2)], \tag{6.72}$$

and

$$X = (\epsilon_1-\epsilon_2)^2+4V^2-(1/4)(\gamma_1+\gamma_2)^2-i[(\epsilon_1-\epsilon_2)(\gamma_1-\gamma_2)+4VA_1A_2]. \quad (6.73)$$

Two simple limiting cases highlight underlying physics. For stable states, $\gamma_{1,2}=0$, we have a standard level repulsion driven by the Hermitian interaction V as it was stressed in our discussion of the level dynamics. This served as a simple driving force behind the mixing of the wave functions and establishing the final "disordered crystal" of the level spectrum and the spacing (or ratio) distribution function. This works also for our states originally in the continuum. A sufficiently strong interaction V repels the states, at some point the lower one acquires negative energy becoming stable. This is a rough image of a possible mechanism responsible for the existence of loosely bound nuclei, for example due to the pairing-type mixing of two-particle states on s- and p-orbitals (in reality the collective degrees of freedom of the core may also influence this picture [187]).

For the degenerate unperturbed states, $\epsilon_{1,2}=\epsilon$, without a direct interaction, $V=0$, the solution predicts two states still degenerate at real energy ϵ but with the widths $\Gamma_1=0$ and $\Gamma_2=\gamma_1+\gamma_2$. This is a limiting case of the superradiance when one resonance accumulates the total continuum coupling. The stable state of the pair still can have real energy higher than the formal threshold giving an example of a "bound" state immersed in the continuum. Such mechanisms were studied in more detail theoretically [49, 50] and experimentally with the use of microwave cavities [183]. From a more general theoretical viewpoint, this is a special case of the so-called set of *exceptional points* [159] known in many applications.

As seen from the structure of our variable X, Eq. (6.73), the real coupling (the parameter V) and the continuum coupling (related to partial decays $\gamma_{1,2}$) act essentially in the opposite way. The coupling through continuum, repelling the widths of two states, effectively moves closer the real parts of energy [49]. In distinction to the GOE picture of non-crossing levels, in the presence of continuum coupling, the real energies can coincide. This was clearly seen in the continuum modeling for the random ensemble (GOE + continuum), Ref. [162]. Roughly speaking, a Hermitian interaction *repels the energies* but, due to the physical mixing, makes their continuum properties similar, *attracts the widths*. Opposite to that, the non-Hermitian decay interaction through common continuum channels *attracts the energies* (actually centers of the resonances) but tries to move the system to superradiance and makes the lifetimes of the states different (*repels the widths*). Always, the resonance positions in the complex plane are not crossing.

At $V^2 = \epsilon_1\epsilon_2$, the lower root of the usual algebraic solution for the stationary two-level problem comes to the zero threshold, after which this state becomes indeed bound as below threshold the decay amplitudes have to vanish. At this critical point, the upper state corresponds to the energy

$$E_+ = \frac{1}{2}\left[\epsilon_1 + \epsilon_2 + \sqrt{(\epsilon_1+\epsilon_2)^2 + \Gamma_+(\Gamma_+ - \gamma_1 - \gamma_2)}\right]. \qquad (6.74)$$

Calculating the imaginary part of this root, we find

$$\Gamma_+(E_+) = \gamma_1(E_+) + \gamma_2(E_+) - \frac{[A_1(E_+)\sqrt{\epsilon_2} - A_2(E_+)\sqrt{\epsilon_1}]^2}{\epsilon_1+\epsilon_2}. \qquad (6.75)$$

Here the energy dependence of the decay amplitudes is critical bringing in the violation of the trace, $\Gamma_+ \neq \gamma_1 + \gamma_2$. In the "normal" quantum problem with energy-independent matrix elements, the trace would be conserved for the real part as it is seen from Eq. (6.74). The trace is restored in an exceptional case when the decay amplitudes grow universally keeping the proportionality,

$$\frac{A_1(E)}{A_2(E)} = \sqrt{\frac{\epsilon_1}{\epsilon_2}}. \qquad (6.76)$$

In the prototypical problem of pairing on s- and p-orbitals, this does not happen as widths of p-states grow faster than those of s-states. This, close to reality, case is discussed in detail in Ref. [240], see Fig. 6.6. Note that, at some conspiracy of parameters, namely

$$V = \left[\frac{A_1A_2(\epsilon_1-\epsilon_2)}{\gamma_1-\gamma_2}\right]_{E=E_-}, \qquad (6.77)$$

we have the vanishing width Γ_2 of the lower state while its real energy E_2 is still embedded in the continuum.

The scattering cross section in such a two-level model (in this case the analogy to ^{11}Li does not work as it would correspond to the unrealistic scattering of a bound neutron pair) can be directly found with the scattering amplitude that has two complex poles, $\mathcal{E}_1$ and $\mathcal{E}_2$, of Eq. (6.71),

$$T(E) = \frac{E(\gamma_1+\gamma_2) - \gamma_1\epsilon_2 - \gamma_2\epsilon_1 - 2VA_1A_2}{(E-\mathcal{E}_1)(E-\mathcal{E}_2)}. \qquad (6.78)$$

In the special case of a pair of degenerate intrinsic levels with no direct Hermitian interaction, the general result reduces to the single Breit–Wigner

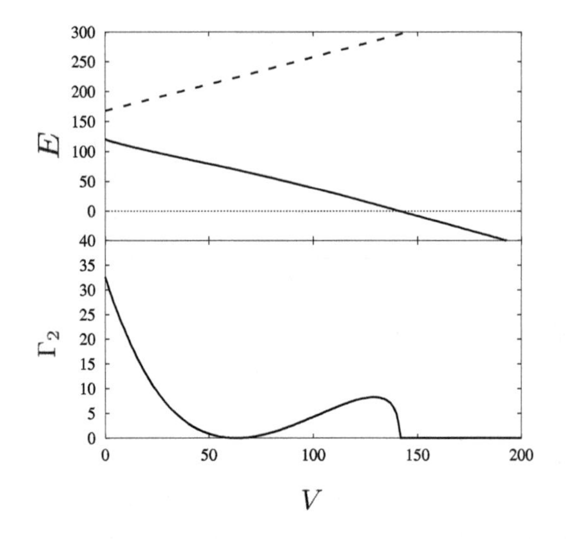

Fig. 6.6. The dynamics of two states, discussed in the two-level model, Ref. [240]. The top panel shows the energies of the two states as a function of the Hermitian off-diagonal interaction strength V; the lower panel shows the width of the lower state Γ_2.

resonance on a superradiating state accumulating the whole width,

$$T(E) = \frac{\gamma_1 + \gamma_2}{E - \epsilon + (i/2)(\gamma_1 + \gamma_2)}, \tag{6.79}$$

The second root is decoupled from the continuum and does not influence the scattering process; the starting widths γ_1, γ_2 in general depend on the running energy E. Figure 6.7 illustrates, for two interesting cases, a non-trivial form of the cross section depending on the sign of the product of the amplitudes $A_1 A_2$.

6.8. Realistic Continuum Shell Model

Here we give an example of the fully realistic continuum shell model based on the united description of the appropriate part of the whole nuclear Hilbert space (discrete and continuum states) using the non-Hermitian effective Hamiltonian as a method of choice [242]. It is clear that this approach, with appropriate modifications, can be applied to other mesoscopic systems outside of nuclear physics, the examples will be given in Sec. 7.

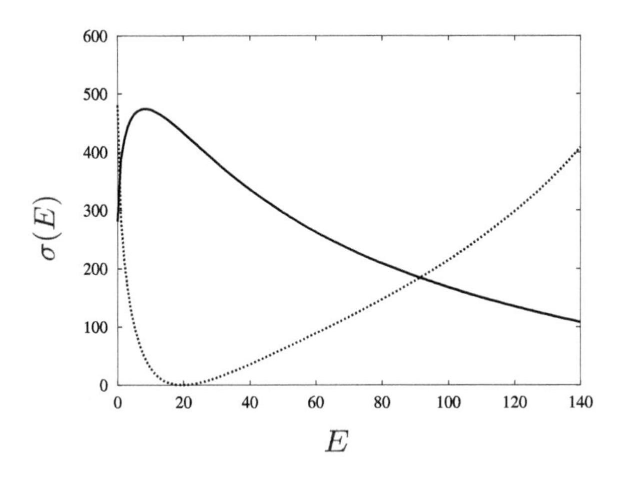

Fig. 6.7. The near-threshold scattering cross section $\sigma(E)$ is shown as a function of energy for a two-level model. The solid curve corresponds to the case of $A_1A_2 > 0$ and the dotted line is for $A_1A_2 < 0$ with all other parameters being the same. Details of the model can be found in Ref. [240].

In nuclear physics, nuclear chemistry, and astrophysics, such a unified approach to bound and unbound states reflects the deep interest to the periodic system of elements as a whole, terrestrial and general abundances of various isotopes, and evolution of the universe. Special attention is attracted by the chains of helium, oxygen and neighboring isotopes serving as a gate to the production of heavier chemical elements [81, 225, 226]. Experimentally, the heaviest particle-stable oxygen isotope is ^{24}O with the half-life of 24 ms. The next even-even isotope ^{26}O is barely particle-unstable decaying mainly by emission of a neutron pair with the half-life of 40 ns, quite long on the nuclear scale with strong interactions.

The continuum shell model in this approach is erected on the foundation of the Feshbach projection method. An exact stationary state $|\alpha\rangle$ with energy E in general has components in both subspaces, $\mathcal{Q}$ (the part of the space independently existing in the absence of continuum coupling) and $\mathcal{P}$ coming as a result of the finite lifetime if the real decay is energetically allowed; this subspace contains the contributions of channels c open at energy E,

$$|\alpha\rangle = \left[\sum_{1} \alpha_1(E)|1\rangle\right]_{\mathcal{Q}} + \left[\sum_{c} \int dE' \, A^c(E', E)|c; E'\rangle\right]_{\mathcal{P}} . \tag{6.80}$$

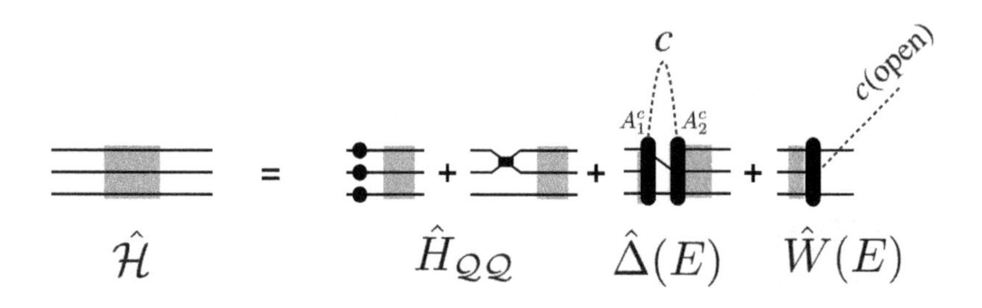

Fig. 6.8. Diagram of the effective Hamiltonian.

We assume the eigenchannel diagonalization in the $\mathcal{P}$-space as discussed in [82],

$$H_{\mathcal{PP}}|c;E\rangle = E|c;E\rangle, \tag{6.81}$$

Then the amplitudes of the continuum admixture in the full wave function (6.80) are

$$\alpha_c(E',E) = \frac{1}{E-E'+i\eta}\sum_1 \alpha_1(E)A_1^{c*}(E',E), \tag{6.82}$$

where $\eta \to +0$. The continuum amplitudes for the channel c depending on the actual energy E of the state and the running variable E' can be defined as the coupling coefficients between the two subspaces at a given energy E,

$$A_1^c(E',E) = \langle 1|H-E|c;E'\rangle. \tag{6.83}$$

The state vectors $|c;E'\rangle$ carry the memory of the threshold energy E_c with amplitudes going to zero at $E' \to E_c$. Finally, the problem is formulated as that of the diagonalization in the intrinsic ($\mathcal{Q}$) space of the effective non-Hermitian Hamiltonian matrix (6.10), that explains a clear physical meaning of the equation, see Fig. 6.8,

$$(\mathcal{H}(E))_{12} = H_{12} + \sum_c \int dE' \, \frac{A_1^c(E',E)A_2^{c*}(E',E)}{E-E'+i\eta}. \tag{6.84}$$

As discussed earlier, the imaginary part $-(i/2)W(E)$ of the effective Hamiltonian (6.84), defined by the singularity in the denominator, is a sum of factorized terms corresponding to the open channels, $E' = E \geq E_c$,

$$W(E) = 2\pi \sum_{c(\text{open})} A_1^c(E,E)A_2^{c*}(E,E). \tag{6.85}$$

The Hermitian part of the total effective Hamiltonian (6.84) includes also the continuum correction $\Delta(E)$ coming from the principal value terms describing the virtual coupling to the decay channels (both, open and closed at given energy E).

We are working with a non-Hermitian Hamiltonian but still in the shell-model space, in distinction to approaches using the direct discretization of the continuum. The eigenvalue problems with such a Hamiltonian require finding the two sets of adjoint eigenvectors,

$$\mathcal{H}|\alpha\rangle = \mathcal{E}_\alpha|\alpha\rangle; \quad \langle\tilde{\alpha}|\mathcal{H} = \mathcal{E}^*_\alpha\langle\tilde{\alpha}|. \tag{6.86}$$

The left and right eigenstates correspond to time-reversed motions; they no longer have to coincide because the $\mathcal{T}$ invariance in the internal space is broken by irreversible decays, while the global symmetry with respect to the direction of time is, however, maintained by the full Hamiltonian that includes the products of reactions. As a result, the left and right eigenstates have the wave functions interrelated by complex conjugation that serves as the time-reversal operation.

Above thresholds, the diagonalization of the matrix (6.84) provides complex energy-dependent eigenvalues

$$\mathcal{E}_\alpha(E) = E_\alpha(E) - \frac{i}{2}\Gamma_\alpha(E), \tag{6.87}$$

where the imaginary part satisfies the Bell–Steinberger relation [29, 209], $\Gamma_\alpha = \langle\tilde{\alpha}|W|\alpha\rangle$. Wide resonances cover broad regions of energy and can be noticeably affected by the energy dependence in this formulation. This may lead to a non-generic and asymmetric shape of a resonance cross section corresponding to the non-exponential decay [181] that makes standard Breit–Wigner or Gaussian parameterizations insufficient. In such cases, the whole concept of individual resonances should be substituted by the energy-dependent cross sections uniquely compared to the experiments.

Practical calculations currently, as a rule, are restricted to one and two-body interactions. The $\mathcal{P}$ subspace is usually limited by the states with only one or two nucleons in the continuum, while the typical shell-model limitations by few valence shells are imposed on the intrinsic Q subspace. The internal interaction H_{QQ}, together with the Hermitian self-energy term Δ included, is simply identified with a standard shell-model Hamiltonian [53, 54]. In principle, the parameters are to be readjusted as functions of running energy which was not done yet. The continuum basis states are taken as antisymmetrized products of internal eigenstates of the residual nucleus and the wave functions of one or two particles in the continuum. In principle, this should be generalized to the bound clusters of emitted particles, starting with the deuterons and alpha particles.

Here we do not show all details of actual calculations referring to the original work [242]. The coupling to the continuum that is absent in the standard shell model requires additional information that, at this stage of development, can be taken from the solution of the scattering problem with the potential created by the rest of the nucleus described by the shell model. Although, from the pure theoretical viewpoint, this is just a working "poor man's" solution, in the relatively light nuclei it is practically sufficient. The complex energies of the resonances are found as poles of the scattering matrix. Special attention should be paid to the near-threshold behavior of the wave functions and scattering amplitudes, with the corresponding energy dependence of single-particle widths $\gamma_\ell \propto \epsilon^{\ell+1/2}$. The $\mathcal{Q}$ and $\mathcal{P}$ subspaces are delineated by the solution of the effective potential created by the shell model. The solution reproduces the lifetimes of the oxygen isotopes and correctly predicts, in agreement with the subsequent experiments, the drip line at ^{24}O with the long-lived isotope ^{26}O.

Even on the level of one-particle decays, this approach goes beyond the simple traditional formulation with the spectroscopic factors which frequently just mask the absence of certain many-body physical arguments. The effects of restructuring, shape modification, change of energetics, etc. are especially sensitive to the thresholds and the boundaries of the drip line.

Here we present one example from [265] that highlights the combined experimental and theoretical (in the framework of the continuum shell model) investigations of the ^{8}B nucleus using P$+^7$Be reaction. The nucleus ^{7}Be has the ground state $3/2^-$ and the low-lying first excited state $1/2^-$ at energy 0.429 MeV, see Fig. 6.9, while the next excited state is much higher, at energy 4.57 MeV. Thus, there are two main decay channels of interest. The proton decay threshold (proton separation energy) in ^{8}B is at 0.137 MeV, correspondingly the threshold for channel that involves an excited state is at 0.567 MeV. Some beta and electromagnetic decay channels are also open but amplitudes for those are by many orders of magnitude smaller.

The upper panel in Fig. 6.10 shows the elastic $p+^7$Be reaction where we start with the ^{7}Be ground state and after scattering end up in the same ground state. The threshold for this ^{7}Be$(p,p)^7$Be elastic channel (proton decay threshold in ^{8}B) at 0.137 MeV is shown by the first vertical dashed

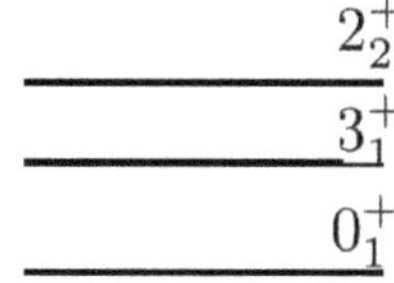

Fig. 6.9. Level scheme of ^{8}B and ^{7}Be in P+^{7}Be channel. Levels are marked with spin and parity, energies are shown in units of MeV relative to ^{8}B ground state.

line. With charged particles, the scattering cross section at low energy is dominated by the Rutherford scattering. The first peak seen here corresponds to a 1^+ resonance. The second broad peak on the top panel is due to a combined contribution from several resonances that is difficult to disentangle. This highlights a typical problem, while additional angular measurements and/or other channels can help along with the theory of the continuum shell model. The lower panel shows the inelastic cross section measured at two different angles. In this case, the excited final state in ^{7}Be is populated; the corresponding $^7\mathrm{Be}(p,p')^7\mathrm{Be}*$ threshold is shown by the second dashed line. Continuous lines in both panels show the result based on a specific shell-model Hamiltonian, while scattered points come from the experiments. The agreement with the data in the elastic channel is very good.

The results for the inelastic processes are sensitive to the position and structure of the 2_2^+ state. The traditional shell model [67] suggest several possible 2^+ states in this energy region with different structure; in the continuum shell-model application, these states lead to a different angular dependence of the inelastic cross section. It is important that dotted curve corresponding to 138° is above the solid line for 129°. In this example, where no adjustable parameters were introduced, the inelastic peak is underestimated by the theory, but further studies and comparison with the R-matrix resolve the discrepancy [9, 161, 192].

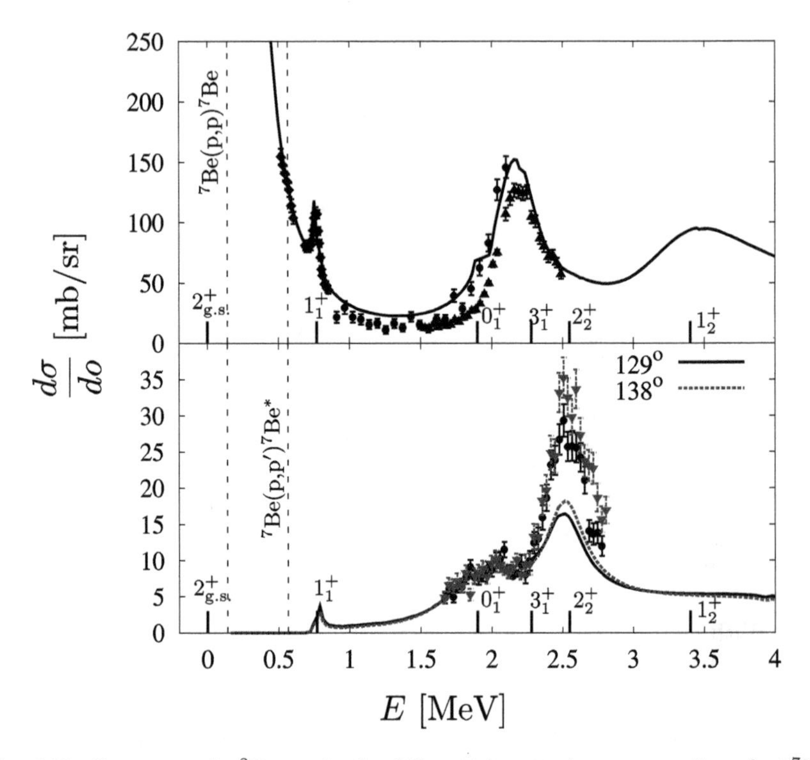

Fig. 6.10. Resonances in ^{8}B seen in the differential scattering cross section of p+^{7}Be reaction. Top panel shows elastic scattering, lower panel shows inelastic scattering populating the excited $1/2^-$ state in ^{7}Be. Points give experimental results [9, 161, 192], curves show the continuum shell model calculations.

Unitarity of the scattering matrix in Eq. (6.20) is an important part of the approach. In the multi-channel scattering, the conservation of the particle number means that the opening of a new channel amounts to a part of the flux being diverted into that channel, so that the cross sections experience discontinuities known as Wigner cusps [27]. Although it is not noticeable in Fig. 6.10, this behavior is present and the elastic cross section (top panel) experiences a minor change at energy 0.567 MeV when the inelastic channel opens. This cusp-type behavior can be much more noticeable in certain cases, such as for s-wave neutral particles where there is no Coulomb and centrifugal barrier; this can serve as a useful tool in studies of unstable systems.

Chapter 7

Quantum Signal Transmission

7.1. Generic Transmission Line

Any physical reaction with participation of a complex nucleus is, in operational sense, a quantum signal transmission through a mesoscopic quantum device that has its own internal degrees of freedom and can transfer the signal, with or without delay and transformation (or distortion). Within such an extremely general understanding, this is just a particular transmission line with its exit as a probabilistic function of the entrance. What is important, this is a quantum device with possible classical measuring procedures at the ends (or at intermediate points) that can provide, under conditions of sufficient statistics, a distribution of numerical results. The science of quantum information [171, 232] deals just with the systems of this type. Below, using simplified examples based on our nuclear experience, we try to illustrate quite general properties of such systems without too detailed specification of actual physical realizations.

As quantum transmission lines can serve in many different application schemes, it is necessary to know their main properties. In this way, we show the simplest constructions that rely on the schematic image of a nuclear reaction as a chain of virtual quantum transitions in and out the system bordered by the entrance and exit channels. This formulation based on the effective non-Hermitian Hamiltonian was first used in Refs. [209, 211] and translated into a more realistic scheme of the signal transmission through a low-dimensional solid state line [58, 59].

A system we start with is modeled as a linear chain of elements (crystal cells, qubits of various material realizations, spins, or individual particles) numbered from $n = 1$ to $n = N$ and connected by internal interactions playing the role of an intermediary for the signal transmission. The ends of the chain are coupled to the continuum channels with the probabilities γ^L

on the left edge ($n = 1$) and γ^R on the right edge ($n = N$). For the closed chain, $\gamma^L = \gamma^R = 0$, we assume the zero boundary conditions at $n = 0$ and $n = N + 1$. The intrinsic Hermitian Hamiltonian H of an autonomous system is described by matrix elements $H_{nn'}$. It can be diagonalized producing N stationary states $|q\rangle$ like the Bloch waves with the quasimomentum q in a periodic chain,

$$|q\rangle = \sum_n |n\rangle\langle n|q\rangle. \tag{7.1}$$

In a simple uniform chain with coupling between the closest neighbors, the quantum number q is similar to the quasimomentum, $\langle n|q\rangle \propto \sin[\pi nq/(N+1)]$, and the energies ϵ_q of a closed system fill in the Bloch zone (the "wave representation"). In a general case, q enumerates the eigenstates of the internal life in a closed system, an analog of the compound nucleus.

In the original "coordinate" representation, the effective non-Hermitian Hamiltonian $\mathcal{H}$ has matrix elements

$$\mathcal{H}_{nn'} = H_{nn'} - \frac{i}{2}\left(\gamma^L\delta_{n1}\delta_{n'1} + \gamma^R\delta_{nN}\delta_{n'N}\right). \tag{7.2}$$

In the intrinsic "wave" representation, the Hamiltonian H is diagonal, while

$$\mathcal{H}_{qq'} = \epsilon_q\delta_{qq'} - \frac{i}{2}\gamma^L\langle q|1\rangle\langle 1|q'\rangle - \frac{i}{2}\gamma^R\langle q|N\rangle\langle N|q'\rangle. \tag{7.3}$$

Similarly to our earlier experience with factorized Hamiltonians, it is easy to diagonalize $\mathcal{H}$. Its quasistationary eigenstates are superpositions of standing waves,

$$|\Psi\rangle = \sum_q X_q|q\rangle, \tag{7.4}$$

with complex energies $\mathcal{E}$ as the eigenvalues of the equation set

$$\mathcal{E}X_q = \sum_{q'} X_{q'}\mathcal{H}_{qq'}. \tag{7.5}$$

For the Hamiltonian (7.3), the amplitudes X_q satisfy

$$X_q = -\frac{i}{2(\mathcal{E} - \epsilon_q)}\left[\gamma^L\langle q|1\rangle Y_1 + \gamma^R\langle q|N\rangle Y_N\right], \tag{7.6}$$

where the effective external couplings of individual states $|q\rangle$ determine the fluxes through the boundaries,

$$Y_1 = \sum_q \langle 1|q\rangle X_q, \quad Y_N = \sum_q \langle N|q\rangle X_q. \tag{7.7}$$

Introducing the propagators between the open channels ($a, b = 1$ or N in this model),

$$P_{ab}(\mathcal{E}) = \sum_q \frac{\langle a|q\rangle\langle q|b\rangle}{\mathcal{E} - \epsilon_q}, \tag{7.8}$$

the amplitudes (7.5) satisfy the algebraic set of coupled equations

$$Y_1\left(1 + \frac{i}{2}\gamma^L P_{11}\right) + Y_N\,\frac{i}{2}\gamma^R P_{1N} = 0, \tag{7.9}$$

$$Y_1\,\frac{i}{2}\gamma^L P_{N1} + Y_N\left(1 + \frac{i}{2}\gamma^R P_{NN}\right) = 0. \tag{7.10}$$

This leads to the roots defined by the secular equation

$$\mathcal{D}(\mathcal{E}) \equiv \left(1 + \frac{i}{2}\gamma^L P_{11}\right)\left(1 + \frac{i}{2}\gamma^R P_{NN}\right) + \frac{\gamma^L\gamma^R}{4}\,P_{1N}P_{N1} = 0. \tag{7.11}$$

Such a formulation is universal for any construction with internal standing waves and two open ends.

In the limit of weak openness, we keep in (7.11) only the terms linear in $\gamma^{L,R}$. Then every standing wave q independently acquires its width, while resonances have long lifetimes and can be labeled by the original waves,

$$\mathcal{E} \to \mathcal{E}_q = \epsilon_q - \frac{i}{2}\Gamma_q. \tag{7.12}$$

The small widths are determined by the overlap of a given mode with the endpoints,

$$\Gamma_q = \gamma^L|\langle 1|q\rangle|^2 + \gamma^R|\langle N|q\rangle|^2. \tag{7.13}$$

If $\Delta\epsilon$ is the band width for the closed system, the extreme super-radiating limit corresponds to $|\mathcal{E}| \gg |\Delta\epsilon|$ when the radiation time is faster than the time of intrinsic propagation. The super-radiating Dicke states move to the ends coupled to the outside world, and the sums (7.8) reduce to

$$P_{ab}(\mathcal{E}) \Rightarrow \frac{1}{\mathcal{E}}\,\delta_{ab}. \tag{7.14}$$

In this limit, the Dicke states are absorbing the whole widths,

$$\mathcal{E}^L = -\frac{i}{2}\gamma^L, \quad \mathcal{E}^R = -\frac{i}{2}\gamma^R. \tag{7.15}$$

We can also find the positions of these resonances on the real energy axis (always close to the center of the band) and corrections to the Dicke limit which include small widths of trapped states, $\propto (\Delta\epsilon)^2/\gamma$.

7.2. Transmission Landscape

Varying the interactions inside the system and its coupling with the outside world, one can regulate the transmitted signal. As discussed in Chapter 6, the process observed from outside the system can be described by the reaction matrix (the hat here refers to operators in the open channel space, just 2×2 in the current model)

$$\hat{R}(\mathcal{E}) = \mathbf{A}^T G(\mathcal{E})\mathbf{A}. \tag{7.16}$$

The vector $\mathbf{A}$ connects internal propagation with the decay channels (only two in the previously discussed model), while $G(\mathcal{E})$ is the propagator for an isolated intrinsic system,

$$G(\mathcal{E}) = \frac{1}{\mathcal{E} - H}. \tag{7.17}$$

The total propagator is

$$\mathcal{G}(\mathcal{E}) = \frac{1}{\mathcal{E} - \mathcal{H}} = G - \frac{i}{2} G\mathbf{A}\,\frac{1}{1 + (i/2)\hat{R}}\,\mathbf{A}^T G; \tag{7.18}$$

this is sometimes called the Woodbury equation.

The total transmission matrix in the channel space, again 2×2 in the simplified scheme of Sec. 7.1, is

$$\hat{T}(\mathcal{E}) = \mathbf{A}^T \mathcal{G}(\mathcal{E})\mathbf{A}, \tag{7.19}$$

It corresponds to an arbitrary number of excursions into the channels and returning back,

$$\hat{T} = \frac{\hat{R}}{1 + (i/2)\hat{R}} \equiv \hat{R}\hat{K}(\hat{R}). \tag{7.20}$$

The full scattering matrix $\hat{S}$ including the free propagation outside the system is obviously unitary at real energy,

$$\hat{S} = 1 - i\hat{T} = \frac{1 - (i/2)\hat{R}}{1 + (i/2)\hat{R}}, \quad \hat{S}^{\dagger}\hat{S} = \hat{1}. \tag{7.21}$$

The resonances are defined as the roots of the determinant (7.11) that can be now written in terms of channel space variables,

$$\mathcal{D}(\mathcal{E}) = \det\left[1 + \frac{i}{2}\hat{R}(\mathcal{E})\right] = 0. \tag{7.22}$$

These roots are complex in general while the physical transmission requires to find $\hat{T}(E)$ at the real energy of an experiment.

In the model of Sec. 7.1, the elements of the $\hat{R}$ matrix, Eq. (7.16), are

$$R^{LL} = \gamma^{L}P_{11}, \quad R^{RR} = \gamma^{R}P_{NN}, \tag{7.23}$$

$$R^{LR} = \sqrt{\gamma^{L}\gamma^{R}}\,P_{1N}, \quad R^{RL} = \sqrt{\gamma^{R}\gamma^{L}}\,P_{N1}. \tag{7.24}$$

The elements of the matrix $\hat{K}$, Eq. (7.20), are equal to

$$K^{LL} = \frac{1 + (i/2)R^{RR}}{\mathcal{D}}, \quad K^{RR} = \frac{1 + (i/2)R^{LL}}{\mathcal{D}}, \tag{7.25}$$

$$K^{LR} = -\frac{i}{2}\frac{R^{LR}}{\mathcal{D}}, \quad K^{RL} = -\frac{i}{2}\frac{R^{RL}}{\mathcal{D}}. \tag{7.26}$$

Finally, this leads to the transmission matrix (7.20),

$$T^{LL} = R^{LL}K^{LL} + R^{LR}K^{RL} = \frac{R^{LL} + (i/2)\det\hat{R}}{\mathcal{D}}, \tag{7.27}$$

$$T^{RR} = R^{RR}K^{RR} + R^{RL}K^{LR} = \frac{R^{RR} + (i/2)\det\hat{R}}{\mathcal{D}}, \tag{7.28}$$

$$T^{LR} = \frac{R^{LR}}{\mathcal{D}}, \quad T^{RL} = \frac{R^{RL}}{\mathcal{D}}. \tag{7.29}$$

If the intrinsic spectrum ϵ_q is found, and the corresponding eigenstates $|q\rangle$ are known in terms of the original localized states $|n\rangle$, the transmission amplitudes are exactly given by the last set of equations as functions of real energy, $\mathcal{E} \to E$.

7.3. Tight-Binding Model

Here we look at the simplest one-dimensional chain with the nearest neighbor couplings. It was first considered from our current transmission viewpoint [209], discussed further in [241], and translated into the realistic solid-state language [58, 59], where it was argued that the super-radiance can actually occur in real applications. The solid-state analog is an array of one-dimensional quantum dots [166, 167, 229]. For the simplest arrangement, we follow Ref. [59] considering a regular sequence of $N+1$ potential barriers of height V and spatial width Δ for $(N-1)$ internal barriers separated by a distance L. The external barriers have spatial widths Δ_L and Δ_R (not to be confused with the energy widths). For translating into a practical language, in units $\hbar^2/(2m) = 1$, where m is the electron mass, all lengths are measured in nanometers, and energies in units of 38 meV; the center of the energy band is $\epsilon_0 = V/2 = 500$.

The transmission intensity through the chain is given by the absolute square $|T^{ab}|^2$ of the corresponding matrix element of the appropriate matrix (7.19). The non-vanishing amplitudes are $A_1^L = \sqrt{\gamma^L}$ and $A_N^R = \sqrt{\gamma^R}$. The sites of the chain are interconnected by the tunneling amplitude Ω that is determined by the geometry and material of the chain. The transmission through the chain from the left to the right at real energy E is characterized by

$$T^{LR}(E) = \frac{\sqrt{\gamma^L \gamma^R}}{\prod_{n=1}^{N}(E - \mathcal{E}_n)}, \tag{7.30}$$

where $\mathcal{E}_n$ are complex eigenvalues of the effective Hamiltonian $\mathcal{H}$. There is a trivial case of a single level with the different tunneling amplitudes $\gamma^L = \gamma$ and $\gamma^R = \gamma/q$ with a certain asymmetry parameter q. Then one can immediately see that the maximum transmission $T^{LR} \equiv T$ corresponds to $E = 0$ when

$$|T|^2 = \frac{4q}{(q+1)^2}, \tag{7.31}$$

the expression obviously symmetric with respect to the change $q \to 1/q$. The maximum transmission equal to one happens only in the symmetric case, $q = 1$.

The situation becomes more interesting if there are at least two intrinsic states. We have already met a similar problem in the context of the continuum shell model [240], see also [5]. Here the effective Hamiltonian for two originally degenerate intrinsic states has off-diagonal coupling Ω and diagonal terms $(-i/2)(\gamma, \gamma/q)$. The transmission at $E = 0$ is given by

$$T = \frac{x}{1 + x^2/4}, \quad x = \frac{\gamma}{\Omega\sqrt{q}}. \tag{7.32}$$

The perfect transmission, $T = 1$, is achieved at $x = 2$; this also provides the maximum of the transmission integrated over the energy band. Moreover, this behavior is actually not sensitive to the number N of sites reaching the maximum at the same value of x as was discussed already in [211], see Fig. 7.1.

In the general case of N identical sites, the results still can be derived analytically [59]. The transmission amplitude in this case is given by

$$T(E) = \frac{2\sqrt{\gamma^L\gamma^R}\,P_-}{1 + i(\gamma^L + \gamma^R)P_+ + \gamma^L\gamma^R(P_-^2 - P_+^2)}. \tag{7.33}$$

Here the energy-dependent sums,

$$P_\pm(E) = \sum_{n=1}^{N} \frac{(\pm)^n}{N+1} \sin^2\varphi_n \frac{1}{E - \epsilon_n}, \tag{7.34}$$

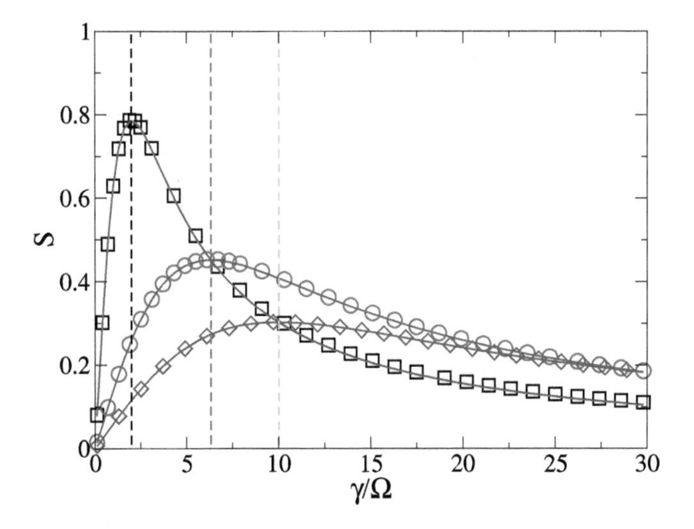

Fig. 7.1. Integrated transmission S for the $N = 100$ system as a function of γ/Ω for different values of the asymmetry parameter $q = 1, 10$, and 25 using circles, squares, and diamonds, respectively. The solid curves show the $N = 2$ case, and vertical dashed lines indicate peak locations where $\gamma/\Omega = 2\sqrt{q}$ and $x = 2$. (Reprinted with permission from Ref. [59]; copyright (2010) by the American Physical Society.)

can be calculated exactly. The internal energies here are

$$\epsilon_n = 2\Omega \cos \varphi_n, \quad \varphi_n = \frac{\pi n}{N+1}. \tag{7.35}$$

With dimensionless energy $\nu = E/(2\Omega)$ and the complex variable $z_n = e^{i\varphi_n}$, it is convenient to double the sum with the phase φ_n running over the entire circle, so that the variables z_n are the roots of

$$z^{2N+2} = 1. \tag{7.36}$$

In these variables,

$$P_\pm(\nu) = \sum_{n=0}^{2N+1} \frac{(\pm)^n}{4\Omega(2N+2)} \frac{(z_n^2 - 1)^2}{z_n(z_n^2 - 2\nu z_n + 1)}. \tag{7.37}$$

The roots of the quadratic equation $z^2 - 2\nu z + 1 = 0$ are $z_\pm = \nu \pm \sqrt{\nu^2 - 1}$, and $z_+ z_- = 1$.

Using the standard summations,

$$\sum_{n=0}^{2N+1} \frac{(\pm)^n}{(2N+2)z_n - z} = -\frac{z^N}{2(z^{N+1} - 1)} \mp \frac{z^N}{2(z^{N+1} + 1)}, \tag{7.38}$$

we come to the results for the required sums,

$$2\Omega P_+(\nu) = \nu - \frac{2(\nu^2 - 1)}{z_+ - z_-} \left(\frac{z_+^{2N+2}}{z_+^{2N+2} - 1} - \frac{z_-^{2N+2}}{z_-^{2N+2} - 1} \right), \tag{7.39}$$

$$-\Omega P_-(\nu) = \frac{\nu^2 - 1}{z_+ - z_-} \left(\frac{z_+^{N+1}}{z_+^{2N+2} - 1} - \frac{z_-^{N+1}}{z_-^{2N+2} - 1} \right), \tag{7.40}$$

The expressions are simplified for the energies inside the band; with $\nu = \cos \beta$,

$$P_+(\nu) = \frac{\sin(N\beta)}{2\Omega \sin[(N+1)\beta]}, \quad P_-(\nu) = -\frac{\sin(\beta)}{2\Omega \sin[(N+1)\beta]}. \tag{7.41}$$

For a general N case, the perfect transmission is found for all resonances at the same condition $x = 2$ along with the maximum of the transmission integrated over the width of the energy band, Fig. 7.2 from [59] shows the ratio of the mean width of $N - 2$ remaining resonances after the two super-radiating state are excluded. The maximum transmission is always in between the two super-radiating transitions. The number of observed resonances changes after the exclusion of each very broad super-radiating

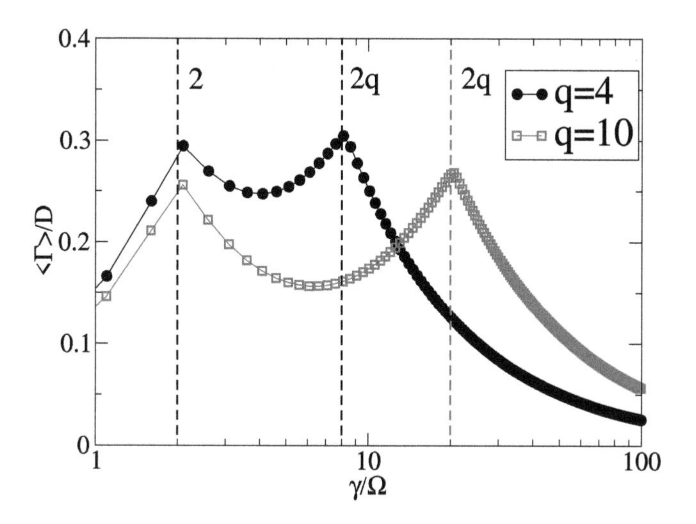

Fig. 7.2. Average width (excluding the two largest widths) as a function of γ/Ω for the system with $N = 100$. The width is shown in the units of level spacing at the center of the band. The two values of the asymmetry parameter, $q = 4$ and 10, are shown. The vertical dashed lines indicate positions of two superradiant transitions, $\gamma/\Omega = 2$ and $\gamma/\Omega = 2q$. (Reprinted with permission from Ref. [59]; copyright (2010) by the American Physical Society.)

states whose positions are exactly given by the same ratios γ/Ω equal to 2 and $2q$ as in the case of two sites.

An important development corresponds to the *disordered* chains. It was established earlier [209, 241] that the superradiance mechanism survives in the presence of disorder being described by the same effective non-Hermitian Hamiltonian. Here the results close up to the very influential development for disordered solid-state systems with Anderson localization [7, 146]. This would lead us rather far from our main topics but it is necessary to stress that the effective Hamiltonian approach allows one to consistently derive many of those practically important results in this fast developing area. We can mention the possible artificial insertion of disordering elements into the transmission line [196] in order to avoid the resonance coupling between the qubits (and switch it on at a necessary moment) and in this way exclude the formation of many-qubit chaotic states. Figure 7.3 shows the transmission through the open Anderson model found with the effective non-Hermitian Hamiltonian. This paper also provides a detailed comparison with the formal random matrix theory. It turns out that the specific features of an open quantum system modify the standard results of that theory.

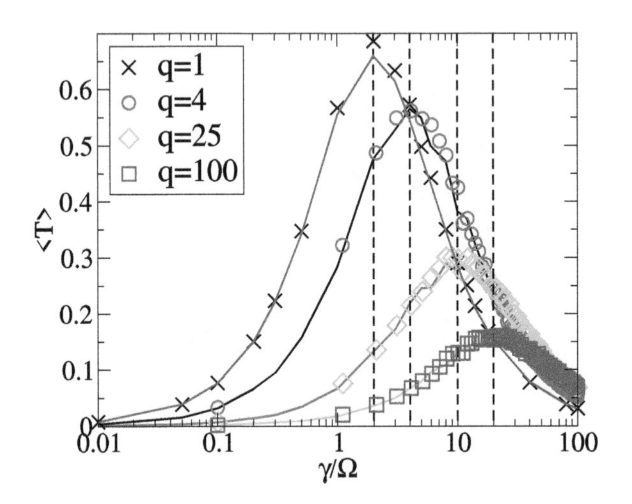

Fig. 7.3. Average transmission in the vicinity of $E = 0$ through a disordered one-dimensional chain of 100 wells with different values of asymmetry parameters. (Reprinted with permission from Ref. [59]; copyright (2010) by the American Physical Society.)

The detailed consideration of the photon propagation through a one-dimensional qubit system based on the same theoretical approach can be found in Refs. [112, 113]. On the way to quantum computing, a system of two qubits was studied in the same formalism, mainly from the viewpoint of the signal propagation and the dependence of the resulting signal on the qubit positions [221]. A more refined scheme [224] helps to construct an environment protected qubit and study the influence of noise.

7.4. Network Propagation

The whole formalism can be extended to configurations of complicated geometry and topology with possible various paths of internal propagation and also to arrangements with many available entrance and exit points. In a sense, this situation might be closer to a realistic nuclear reaction with its complicated Hilbert space of intrinsic many-body quantum states (competing quantum trajectories) and the presence of a network of open and closed reaction channels.

A two-dimensional open model [59] consists of a rectangular set of wires going in parallel through a lattice with a number of external entrance channel points on the left-hand side and, maybe different, number of exit channel points on the right-hand side. In the most general case, each vertex can have intrinsic quantum levels, and the model can be generalized to greater dimensions. The signal propagates through the neighboring sites. The total conductance as a sum of the squared amplitudes $|T(E)|^2$ for all combinations of the left (entrance) and right (exit) channels has the maximum value at the same condition $x = 2$ as earlier; the detailed behavior of the average transmission is shown in Fig. 7.4.

A different geometry of the network, the nest or the star circuit, was studied in [268]. Such a system is characterized by the presence of the central site and $M > 2$ branches a of N_a cites coupled at the center and capable to connect with the outside world through the ends. If we label the consecutive sites in the branch a by n running from 1 to N_a, we have the coupling amplitudes v_a between the neighbors in the branch a and $t_a v_a$ between this branch and the common center labeled by zero, see Fig. 7.5. The stationary wave function of the whole star graph is described by the central amplitude C_0 and the set of $\sum_{a=1}^{M} N_a$ amplitudes C_n^a. Assuming real and equal transition amplitudes within a given branch, the obvious equations can be written for the stationary state of energy E of the complex

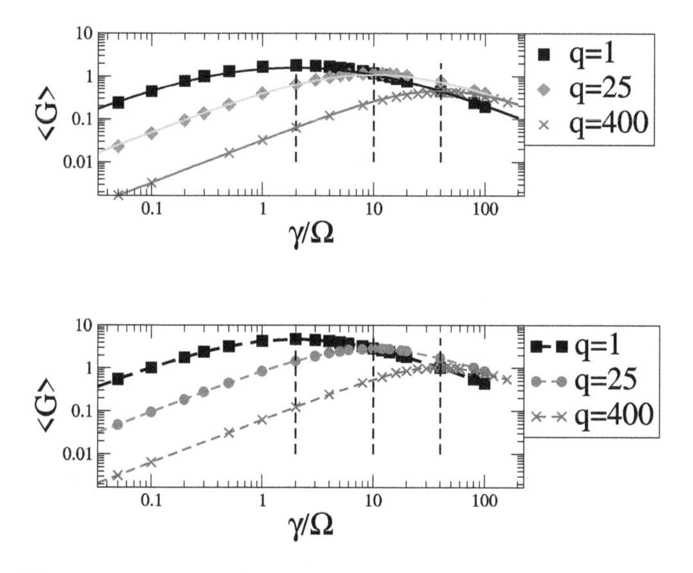

Fig. 7.4. The average dimensionless conductance as a function of the coupling strength. In the upper panel the 10×100 geometry is shown, the lower panel shows 30×30 geometry. Results from numerical studies are shown by squares, diamonds, and crosses for different values of q as marked. Continuous curves are results of random matrix theory. Dashed grid lines indicate critical values $\gamma/\Omega = 2q$. (Reprinted with permission from Ref. [59]; copyright (2010) by the American Physical Society.)

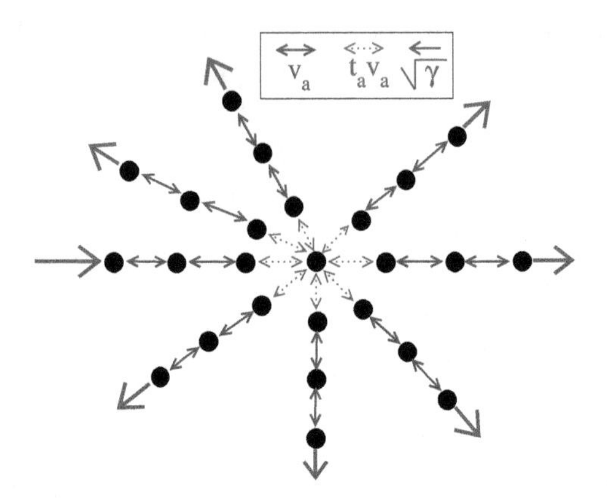

Fig. 7.5. The M-branch circuit with the coupling at the origin. (Reprinted with permission from Ref. [268]; copyright (2012) by the American Physical Society.)

as a whole: propagation along every branch,

$$v_a(C_{n-1}^a + C_{n+1}^a) = EC_n^a, \quad 2 \leq n \leq N_a; \tag{7.42}$$

communication with the center,

$$v_a C_2^a + t_a v_a C_0 = EC_1^a; \tag{7.43}$$

and coupling between the branches through the center,

$$\sum_{a=1}^{M} t_a v_a C_1^a = EC_0. \tag{7.44}$$

For the isolated system, the boundary conditions can be taken as $C_{N_a+1}^a = 0$.

Among possible modes, the most interesting solutions for the closed system correspond to $C_0 \neq 0$; for simplicity one can take identical rays, $v_a = v$ and $N_a = N$. For each individual decoupled branch, we would have the Bloch band of the width $2v$ with energies ϵ depending on the parameter E that defines the position of the band on the energy scale,

$$\epsilon_{\pm}(E) = -E \pm \sqrt{E^2 - 4v^2}. \tag{7.45}$$

The spectrum of the eigenvalues E within the band $|E| < 2v$ is given by the roots of the equation

$$f_+(E)\epsilon_-^N - f_-(E)\epsilon_+^N = 0, \tag{7.46}$$

where

$$f_{\pm}(E) = E^2 - E\epsilon_{\pm} - v^2 Q, \quad Q = \sum_{a=1}^{M} t_a^2. \tag{7.47}$$

The detailed analysis shows that there are $MN - 1$ states within the energy band $|E| < 2v$ (they are spread over all branches) and two additional states with $|E| > 2v$ outside this band; for $Q \gg 1$, their positions are $\pm v\sqrt{Q}$ (this generalizes the result earlier derived in [69]).

For our purpose, the important consideration is that of the properties of an open system. According to our method, we add an anti-Hermitian part and come to

$$\mathcal{H} = H - \frac{i}{2}\gamma W, \tag{7.48}$$

where, in the simplest version, the only physical parameter is γ, while the operator W describes M equiprobable channels governed by the last cells $|a, N\rangle$ of the branches a,

$$W = \sum_{a=1}^{M} |a, N\rangle\langle a, N|. \tag{7.49}$$

At weak continuum coupling, the small widths of the resonances are proportional to γ being naturally defined by the overlaps of originally localized states with the edges. With growing γ, the system demonstrates the transition to superradiance when the formally calculated average width reaches the level of the mean level spacing. After that, the widths of M states (by a number of branches) continue to grow with γ while the remaining states have decreasing widths. The picture of Fig. 7.6 for a graph with $M = 4$ shows the original growth of the mean width of remaining $M(N+1) - M$ states for $N = 70$ with growth of γ. At the critical value $\gamma = 2v$, the sharp transition takes place when this width starts going down and returning to the weak decay while the whole strength is transmitted into M super-radiating states. This story is fully analogous to the early discussed behavior in the simplified version of the nuclear shell model, Fig. 6.2. The observed behavior is qualitatively universal being independent of specific numerical parameters of the graph.

The above-mentioned quasibound resonances with energies $\pm v\sqrt{Q}$ remain quite stable in the process of transition to superradiance. For equal

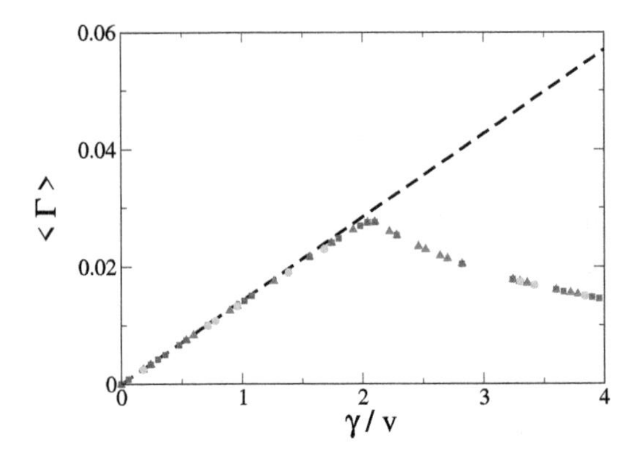

Fig. 7.6. The average width for various models with $M = 4$ but with different number of sites and different couplings. The dashed line provides average over all widths; the symbols show calculations that exclude superradiant states. (Reprinted with permission from Ref. [268]; copyright (2012) by the American Physical Society.)

branches, the transmission, $a \neq b$, is given by a simple expression,

$$T_{a \neq b} = \left(\frac{2t_a t_b}{Q} \right)^2 . \tag{7.50}$$

This result indicates the perspectives of using such schemes for the controllable remittance of the signal. The practically important conclusion is that the special states localized at the center of the open graph become narrow resonances with a large lifetime.

References

[1] C. Abel *et al.*, *Phys. Rev. Lett.* **124**, 081803 (2020).
[2] D. Agassi, H.A. Weidenmüller, and G. Mantzouranis, *Phys. Rep.* **22**, 145 (1975).
[3] V.P. Alfimenkov *et al.*, *Nucl. Phys. A* **398**, 93 (1983).
[4] Y. Alhassid, G.F. Bertsch, S. Liu, and H. Nakada, *Phys. Rev. Lett.* **84**, 4313 (2000).
[5] Y. Alhassid, H.A. Weidenmüller, and A. Wobst, *Phys. Rev. B* **72**, 045318 (2005).
[6] Y. Alhassid, L. Fang, and H. Nakada, *Phys. Rev. Lett.* **101**, 082501 (2008).
[7] P.W. Anderson, *Phys. Rev.* **109**, 1492 (1958).
[8] A.V. Andreev, V.I. Emel'yanov, and Yu.A. Il'inski, *Cooperative Effects in Optics* (IOP, Bristol, 1993).
[9] C. Angulino *et al.*, *Nucl. Phys. A* **716**, 211 (2003).
[10] J.R. Armstrong, S. Åberg, S.M. Reifmann, and V.G. Zelevinsky, *Phys. Rev. E* **86**, 066204 (2012).
[11] V.I. Arnold, *Mathematical Methods of Classical Mechanics* 2nd edn, (Springer, 1989).
[12] P. Arve, *Phys. Rev. A* **44**, 6920 (1991).
[13] Y.Y. Atas, E. Bogomolny, O. Giraud, and G. Roux, *Phys. Rev. Lett.* **110**, 084101 (2013).
[14] Y.Y. Atas, E. Bogomolny, O. Giraud, P. Vivo, and E. Vivo, *J. Phys. A: Math. Theor.* **46**, 355204 (2013).
[15] N. Auerbach, *Phys. Rep.* **98**, 273 (1983).
[16] N. Auerbach and V. Zelevinsky, *Nucl. Phys. A* **781**, 67 (2007).
[17] N. Auerbach and V. Zelevinsky, *J. Phys. G: Nucl. Part. Phys.* **35**, (2008) 093101.
[18] N. Auerbach and V. Zelevinsky, *Rep. Prog. Phys.* **74**, 106301 (2011).
[19] N. Auerbach and V. Zelevinsky, *Phys. Rev. C* **90**, 034315 (2014).
[20] N. Auerbach, V. Zelevinsky, and B. M. Loc, in Weak interactions and nuclear structure, *Nuclear Structure Physics*, A. Shukla and S.K. Patra (eds.), Taylor & Francis Group, 2021, pp. 333–390.
[21] A.V. Avdeenkov and S.P. Kamerdzhiev, *Phys. of At. Nucl.* **72**, 1332 (2009).

[22] P. Axel, *Phys. Rev.* **126**, 671 (1962).
[23] A.B. Balantekin, A. Gouvea, and B. Kayser, *Phys. Lett. B* **789**, 488 (2019).
[24] M. Barbui *et al.*, *Phys. Rev. C* **98**, 044601 (2018).
[25] J. Bardeen, L.N. Cooper, and J.R. Schriffer, *Phys. Rev.* **106**, 162 (1957); **108**, 1175 (1957).
[26] B.R. Barrett, R.F. Casten, J.N. Ginocchio, T. Seligman, and H.A. Weidenmuller, *Phys. Rev. C* **45**, 417 (1992).
[27] A.I. Baz', Ya.B. Zeldovich, and A.M. Perelomov, in Scattering, reactions, and decay in *Non-Relativistic Quantum Mechanics*, (Jerusalem, 1969) (translated from the Russian edition. Science, Moscow, 1966).
[28] C.W.J. Beenakker, *Rev. Mod. Phys.* **69**, 731 (1997).
[29] J.S. Bell and J. Steinberger, in *Oxford Int. Conf. Elementary Particles Proce.*, T.R. Walsh, A.E. Taylor, R.G. Moorhouse and B. Southworth (eds.), (Rutheford High Energy Lab., Chilton, Didicot 1966), p. 195.
[30] S.T. Belyaev, *Kgl. Dansk. Vid. Selsk. Mat.-Fys. Medd.* **31**, No. 11 (1959).
[31] S.T. Belyaev and V.G. Zelevinsky, *Nucl. Phys.* **39**, 582 (1962).
[32] S.T. Belyaev and V.G. Zelevinsky, *Yad. Phys.* **11**, 741 (1970) [*Sov. J. Nucl. Phys.* **11**, 416 (1970)].
[33] M. Bender, P.-H. Heenen, and P.-G. Reinhard, *Rev. Mod. Phys.* **75**, 121 (2003).
[34] M.G. Benedict (ed.), *Super-radiance: Multiatomoc Coherent Emission* (Taylor & Francis, New York, 1996).
[35] K. Bennaceur, F. Nowacki, J. Okolowicz, and M. Ploszajczak, *Nucl. Phys. A* **651**, 289 (1999).
[36] K. Bennaceur, F. Nowacki, J. Okolowicz, and M. Ploszajczak, *Nucl. Phys. A* **671**, 203 (2000).
[37] T. Berggren, *Nucl. Phys. A* **109**, 265 (1968).
[38] C.A. Bertulani and V. Zelevinsky, *Nucl. Phys. A* **568**, 931 (1994).
[39] H. Bethe, *Phys. Rev.* **50**, 332, 977 (1936).
[40] O. Bohigas, M.J. Giannoni, and C. Schmit, *Phys. Rev. Lett.* **52**, 1 (1984).
[41] N. Bohr, *Nature* **137**, 344, 351 (1936).
[42] A. Bohr, B.R. Mottelson, and D. Pines, *Phys. Rev.* **110**, 936 (1958).
[43] A. Bohr and B. Mottelson, *Nuclear Structure* (Benjamin, New York, 1969).
[44] F. Borgonovi, F.M. Izrailev, L.F. Santos, and V.G. Zelevinsky, *Phys. Rep.* **626**, 1 (2016).
[45] F. Borgonovi, F.M. Izrailev, and L.F. Santos, *Phys. Rev. E* **99**, 010101 (2019).
[46] P.F. Bortignon, A. Bracco, D. Brink, and R.A. Broglia, *Phys. Rev. Lett.* **67**, 3360 (1991).
[47] P.F. Bortignon, A. Bracco, and R.A. Broglia, *Giant Resonances* (Harwood Academic Publishers, Amsterdam, 1998).
[48] J.D. Bowman, G.T. Garvey, M.B. Johnson, and G.F. Mitchell, *Ann. Rev. Nucl. Sci.* **43**, 829 (1993).
[49] P. von Brentano, *Phys. Rep.* **264**, 57 (1996).
[50] P. von Brentano and M. Philipp, *Phys. Lett. B* **454**, 171 (1999).

[51] T.A. Brody, J. Flores, J.B. French, P.A. Mello, A. Pandey, and S.S.M. Wong, *Rev. Mod. Phys.* **53**, 385 (1981).
[52] R.A. Broglia and V. Zelevinsky (eds.) *Fifty Years of Nuclear BCS* (World Scientific, Singapore, 2013).
[53] B.A. Brown and B.H. Wildenthal, *Ann. Rev. Nucl. Part. Sci.* **38**, 29 (1988).
[54] B.A. Brown, *Prog. Part. Nucl. Phys.* **47**, 517 (2001).
[55] B.A. Brown and W.A. Richter, *Phys. Rev. C* **74**, 034315 (2006).
[56] P. Carlos *et al.*, *Nucl. Phys. A* **225**, 171 (1974).
[57] G. Casati, I. Guarneri, F. Izrailev, and R. Scharf, *Phys. Rev. Lett.* **64**, 5 (1990).
[58] G.L. Celardo and L. Kaplan, *Phys. Rev. B* **79**, 155108 (2009).
[59] G.L. Celardo, A.M. Smith, S. Sorathia, V.G. Zelevinsky, R.A. Sen'kov, and L. Kaplan, *Phys. Rev. B* **82**, 165437 (2010).
[60] S. Chandrasekhar, *Rev. Mod. Phys.* **15**, 1 (1943).
[61] H.-T.P. Chau, N.A. Smirnova, and P. Van Isacker, *J. Phys. A* **35**, L199 (2002).
[62] N.D. Chavda, Pramana *J. Phys.* **84**, 309 (2015).
[63] N.C. Chávez, F. Mattiotti, J.A. Méndez-Bermúdez, F. Borgonovi, and G.L. Celardo, *Eur. Phys. J.* B**92**, 144 (2019).
[64] B.V. Chirikov, *Phys. Rep.* **52**, 263 (1979).
[65] B.V. Chirikov, *Quantum Mechanics in Chaos*, and Les Houches Lecture Series, M.-J. Giannoni, A. Voros, and J. Zinn-Justin, (eds.) Elsevier, Amsterdam, No. 52, (1991), p. 443.
[66] B.V. Chirikov, *The Problem of Quantum Chaos*, Lecture Notes in Physics, Vol. 411 (Springer, Berlin, 1992) (DOI:10.1007/3-540-56253-2).
[67] S. Cohen and D. Kurath, *Nucl. Phys. A* **73**, 1 (1965).
[68] B.E. Crawford *et al.*, *Phys. Rev. C* **58**, 1225 (1998).
[69] J.L. D'Amato, H.M. Pastawski, and J.F. Weiss, *Phys. Rev. B* **39**, 3554 (1989).
[70] G.V. Danilyan, *Sov. Phys. Usp.* **23**, 323 (1980).
[71] F.F. Deppisch, *A Modern Introduction to Neutrino Physics* (Morgan and Claypool, 2019).
[72] P. Descouvemont and D. Baye, *Rep. Prog. Phys.* **73**, 036301 (2010).
[73] J.M. Deutsch, arXiv:1805.01616v1.
[74] J.M. Deutsch, *Phys. Rev. A* **43**, 2046 (1991).
[75] R.H. Dicke, *Phys. Rev.* **93**, 99 (1954).
[76] V.F. Dmitriev *et al.*, *Nucl. Phys. A* **464**, 237 (1987).
[77] L. Durand, *Phys. Rev. D* **14**, 3174 (1976).
[78] F.J. Dyson, *J. Math. Phys.* **3**, 140, 157, 166, 1199 (1962).
[79] J.-P. Eckmann and D. Ruelle, *Rev. Mod. Phys.* **57**, 617 (1985).
[80] J.B. Ehrman, *Phys. Rev.* **81**, 412 (1951).
[81] J. Emsley, *Oxygen. Nature's Building Blocks: An A–Z Guide to the Elements* (Oxford University Press, Oxford, 2001).
[82] C.A. Engelbrecht and H.A. Weidenmueller, *Phys. Rev. C* **8**, 859 (1973).
[83] E. Epelbaum, A. Nogga, W. Glöckle, H. Kamada, Ulf-G. Meissner, and H. Witala, *Phys. Rev. C* **66**, 064001 (2002).

[84] E. Epelbaum, *Progre. Parti. Nucl. Phys.* **67**, 343 (2012).
[85] T. Ericson, *Adv. Phys.* **9**, 425 (1960).
[86] U. Fano, *Phys. Rev.* **124**, 1866 (1961).
[87] H. Feshbach, *Ann. Phys. (N.Y.)* **5**, 357 (1958).
[88] H. Feshbach, *Ann. Phys. (N.Y.)* **19**, 287 (1962).
[89] H. Feshbach, A. Kerman, and S. Koonin, *Ann. Phys.* **125**, 429 (1980).
[90] V.V. Flambaum, A.A. Gribakina, G.F. Gribakin, and M.G. Kozlov, *Phys. Rev. A* **50**, 267 (1994).
[91] V.V. Flambaum and V.G. Zelevinsky, *Phys. Lett. B* **350**, 1 (1995).
[92] V.V. Flambaum and F.M. Izrailev, *Phys. Rev. E* **56**, 5144 (1997).
[93] V.V. Flambaum and V.G. Zelevinsky, *Phys. Rev. C* **68**, 035502 (2003).
[94] N. Frazier, B.A. Brown and V. Zelevinsky, *Phys. Rev. C* **54**, 1665 (1996). Reprinted in *Spectral Distributions in Nuclei and Statistical Spectroscopy*, V.K.B. Kota and R.U. Haq, (eds.), (World Scientific, Singapore, 2010) p. 557.
[95] J.B. French and K.F. Ratcliff, *Phys. Rev. C* **3**, 94 (1971).
[96] Ya.I. Frenkel, *Izvestia of Academy of Sciences USSR*, **1–2**, 233 (1938).
[97] J.J. Gaardhøje, *Ann. Rev. Nucl. Part. Sci.* **42**, 483 (1992).
[98] J.B. Garg, J. Rainwater, J.S. Petersen, and W.W. Havens, Jr., *Phys. Rev. B* **134**, 985 (1964).
[99] J.D. Garrett, G.B. Hagemann, and B. Herskind, *Ann. Rev. Nucl. Part. Sci.* **36**, 419 (1986).
[100] E. Garrido, P. Sarriguren, E. Moya de Guerra, U. Lombardo, P. Schuck, and H.J. Schulze, *Phys. Rev. C* **63**, 037304 (2001).
[101] A. Gilbert, F.S. Chen, and A.G.W. Cameron, *Can. J. Phys.* **43**, 1248 (1965).
[102] A. Gilbert and A.G.W. Cameron, *Can. J. Phys.* **43**, 1446 (1965).
[103] J. Ginibre, *J. Math. Phys.* **6**, 440 (1965).
[104] S. Giorgini, L.P. Pitaevskii, and S. Stringari, *Rev. Mod. Phys.* **80**, 1215 (2008).
[105] M. Goeppert Mayer and J.H.D. Jensen, *Elementary Theory of Nuclear Shell Structure* (1955).
[106] J. Goldstone, A. Salam, and S. Weinberg, *Phys. Rev.* **127**, 965 (1962).
[107] M.G. Gomez, K. Kar, V.K.B. Kota, J. Retamosa, and R. Sahu, *Phys. Rev. C* **64**, 034305 (2001).
[108] A.L. Goodman, *Phys. Rev. C* **63**, 044325 (2001).
[109] S. Goriely, S. Hilaire, and A.J. Koning, *Phys. Rev. C* **78**, 064307 (2008).
[110] S. Goriely, S. Hilaire, A.J. Koning, M. Sin, and R. Capote, *Phys. Rev. C* **79**, 024612 (2009).
[111] H.-D. Gräf, H.L. Harney, H. Lengeler, C.H. Lewenkopf, C. Rangacharyulu, A. Richter, P. Schardt, and H.A. Weidenmüller, *Phys. Rev. Lett.* **69**, 1296 (1992).
[112] Ya.S. Greenberg, C. Merrigan, A. Tayebi, and V. Zelevinsky, *Eur. Phys. J. B* **86**, 368 (2013).
[113] Ya.S. Greenberg and A.A. Shtygashev, *Phys. Rev. A* **92**, 063835 (2015).
[114] J.-z. Gu and H.A. Weidenmüller, *Nucl. Phys. A* **660**, 197 (1999).

[115] I.I. Gurevich and M.I. Pevzner, *JETP* **31**, 162 (1956) [English **4**, 278 (1957)].
[116] M. Guttormsen *et al.*, *Phys. Rev. C* **88**, 024307 (2013).
[117] M. Guttormsen et al., *Phys. Rev. C* **89**, 014302 (2014).
[118] F. Haake, *Quantum Signatures of Chaos* (Springer, New York, 1991).
[119] C.A. Hampel, ed. *The Encyclopedia of the Chemical Elements* (N.Y. Reinhold Book Corporation, 1968).
[120] M.N. Harakeh and A. van der Woude, *Giant Resonances* (Oxford University Press, 2001).
[121] H.L. Harney, A. Richter, and H.A. Weidenmueller, *Rev. Mod. Phys.* **58**, 607 (1986).
[122] M. Horoi, J. Kaiser, and V. Zelevinsky, *Phys. Rev. C* **67**, 054309 (2003).
[123] M. Horoi, M. Ghita, V. Zelevinsky, *Phys. Rev. C* **69**, 041307(R) (2004).
[124] M. Horoi and V. Zelevinsky, *Phys. Rev. Lett.* **98**, 262503 (2007).
[125] M. Horoi and V. Zelevinsky, *Phys. Rev. C* **75**, 054303 (2007).
[126] M. Horoi and V. Zelevinsky, *Phys. Rev. C* **81**, 034306 (2010).
[127] F. Iachello and A. Arima, *The Interacting Boson Model* (Cambridge University Press, 2006).
[128] H. Iwasaki *et al.*, *Phys. Lett. B* **522**, 227 (2001).
[129] F.M. Izrailev, *Phys. Rep.* **196**, 299 (1990).
[130] R. Kaiser, C. Westbrook, and F. David (eds.), *Coherent atomic matter waves*, NATO Advanced Study Institute (Springer, Berlin, 2001).
[131] L. Kaplan and E.J. Heller, *Ann. Phys.* **264**, 171 (1998).
[132] S. Karampagia, R.A. Sen'kov, and V. Zelevinsky, *Atom Data Nucl. Data Tables* **120**, 1 (2018).
[133] S. Karampagia, V. Zelevinsky, and J. Spitler, *Nucl. Phys. A* **1023**, 122453 (2022).
[134] P. Kleinwachter and I. Rotter, *Phys. Rev. C* **32**, 1742 (1985).
[135] A. Koetzle *et al.*, *Nucl. Instrum. Methods A* **440**, 750 (2000).
[136] W. Kohn and L.J. Sham, *Phys. Rev. A* **140** (1133).
[137] V.K.B. Kota, *Embedded Random Matrix Ensembles in Quantum Physics*, Lecture Notes in Physics, Vol. 884 (Springer, Switzerland, 2014).
[138] V.K.B. Kota and N.D. Chavda, *Int. J. Mod. Phys. E* **27**, 183001 (2018).
[139] M. Kus, M. Lewenstein, and F. Haake, *Phys. Rev. A* **44**, 2800 (1991).
[140] D. Kusnezov, B.A. Brown, and V. Zelevinsky, *Phys. Lett. B* **385**, 5 (1996).
[141] L.D. Landau, *JETP* **7**, 819 (1937) [*Phys. Zs. Sowjet* **11**,556 (1937)].
[142] L.D. Landau and Ya. Smorodinsky, *Lectures on Nuclear Theory* (Springer, New York, 1959).
[143] A.M. Lane and R.G. Thomas, *Rev. Mod. Phys.* **30**, 257 (1958).
[144] K.D. Launey, *Emergent Phenomena in Atomic Nuclei from Large-Scale Modeling* (World Scientific, Singapore, 2017).
[145] B. Lauritzen, P.F. Bortignon, R.A. Broglia, and V.G. Zelevinsky, *Phys. Rev. Lett.* **74**, 5194 (1995).
[146] P.A. Lee and T.V. Ramakrishnan, *Rev. Mod. Phys.* **57**, 287 (1985).
[147] N. Lehmann, D. Saher, V.V. Sokolov, and H.-J. Sommers, *Nucl. Phys. A* **582**, 223 (1995).

[148] C.H. Lewenkopf and V.G. Zelevinsky, *Nucl. Phys. A* **569**, 183c (1994).
[149] S.M. Liddick *et al.*, *Phys. Rev. Lett.* **97**, 082501 (2006).
[150] I.M. Lifshits, S.A. Gredeskul, and L.A. Pastur, *Introduction to the Theory of Disordered Systems* (Wiley, New York, 1988).
[151] A. Likar, *Nucl. Phys. A* **598**, 235 (1996).
[152] A.M. Lyapunov, *The General Problem of the Stability of Motion* (Taylor & Francis, London, 1992).
[153] R. Machleidt and D.R. Entem, *Phys. Rep.* **503**, 1 (2011).
[154] C. Mahaux and H.A. Weidenmueller, *Shell Model Approach to Nuclear Reactions* (North-Holland, Amsterdam, 1969).
[155] M.L. Mehta, *Random Matrices*, 3rd edn. (Elsevier, Amsterdam, 2004).
[156] N. Michel and M. Ploszajczak, *Gamow Shell Model* (Springer Nature, 2020).
[157] A.B. Migdal, *Theory of Finite Fermi Systems and Applications to Atomic Nuclei*, translated from the Russian by S. Chomet (Interscience (Wiley), New York, 1967).
[158] S.L. Miller *et al.*, *Phys. Rev. C* **100**, 014302 (2019).
[159] M.-A. Miri and A. Alu, *Science* **363**, 42 (2019).
[160] G.E. Mitchell, J.D. Bowman, and H.A. Weidenmüller, *Rev. Mod. Phys.* **71**, 445 (1999).
[161] J.P. Mitchell *et al.*, *Phys. Rev. C* **87**, 054617 (2013).
[162] S. Mizutori and V.G. Zelevinsky, Z. *Phys. A* **346**, 1 (1993).
[163] L.G. Moretto, *Nucl. Phys. A* **243**, 77 (1975).
[164] L.G. Moretto, A.C. Larsen, M. Guttormsen, and S. Siem, *AIP Conf. Proc.* **1681**, 040011 (2015).
[165] T. Mori, T.N. Ikeda, E. Kaminishi, and M. Ueda, *J. Phys. B* **51**, 112001 (2018).
[166] G.V. Morozov, D.W.L. Sprung, and J. Martorell, *J. Phys. D* **35**, 2091, (2002).
[167] G.V. Morozov, D.W.L. Sprung, and J. Martorell, *J. Phys. D* **35**, 3052 (2002).
[168] W.F. Mueller *et al.*, *Phys. Rev. C* **73**, 014316 (2006).
[169] D. Mulhall, A. Volya and V. Zelevinsky, *Nucl. Phys. A* **682** (2001) 229c.
[170] D. Mulhall, Z. Huard, and V. Zelevinsky, *Phys. Rev. C* **76**, 064611 (2007).
[171] M.A. Nielsen and I.L. Chuang, *Quantum Computation and Quantum Information* (Cambridge University Press, 2010).
[172] S.G. Nilsson, Kgl. Dansk. Vid. Selsk. *Mat.-Fys. Medd.* **29**, 16 (1955).
[173] P. Nozieres and D. Pines, *Theory of Quantum Liquids* (Avalon Publishing, 1999).
[174] V. Oganesyan and D.A. Huse, *Phys. Rev. B* **75**, 155111 (2007).
[175] M. Ohya and D. Petz, *Quantum Entropy and Its Use* (Springer, Berlin, 1993).
[176] F. von Oppen, *Phys. Rev. Lett.* **73**, 798 (1993).
[177] A. Pandey, *Ann. Phys.* **119**, 170 (1979).
[178] S.P. Pandya, *Phys. Rev.* **103**, 956 (1956).
[179] G. Parisi, *Statistical Field Theory* (Addison-Wesley, 1988).

[180] I.C. Percival, *J. Phys. B* **6**, L229 (1973).
[181] M. Peshkin, A. Volya and V. Zelevinsky, *Europhys. Lett.* **107**, 40001 (2014).
[182] Y.B. Pesin, *Russian Math. Surv.* **32**, 55 (1977).
[183] M. Philipp, P. von Brentano, G. Pascovici, and A. Richter, *Phys. Rev. E* **62**, 1922 (2000).
[184] S.C. Pieper, V.R. Pandharipande, R.B. Wiringa, and J. Carlson, *Phys. Rev. C* **64**, 014001 (2001).
[185] D.N. Poenaru and W. Greiner, Lecture Notes in Physics, Vol. 818, (Springer, Berlin. 2011), p. 1.
[186] C.E. Porter, *Statistical Theories of Spectra: Fluctuations* (Academic Press, New York, 1965).
[187] G. Potel, F. Barranco, E. Vigezzi, and R.A. Broglia, *Phys. Rev. Lett.* **105**, 172502 (2010).
[188] G. Racah, *Phys. Rev.* **62**, 438 (1942).
[189] K.F. Ratcliff, *Phys. Rev. C* **3**, 117 (1971).
[190] J. Reichl, *Europhys. Lett.* **6**, 669 (1988).
[191] R.M. Rockmore, *Phys. Rev.* **116**, 469 (1959).
[192] G.V. Rogachev *et al.*, *Phys. Rev. C* **64**, 616011 (2001).
[193] W. Rossmann, *Lie Groups: An Introduction Through Linear Groups* (Oxford University Press, 2002).
[194] I. Rotter, *Phys. Rev. E* **64**, 036213 (2001).
[195] M. Salaris and S. Cassisi, *Evolution of Stars and Stellar Populations* (Wiley New York, 2005).
[196] L.F. Santos, M.I. Dykman, M. Shapiro, and F.M. Izrailev, *Phys. Rev. A* **71**, 012317 (2005).
[197] W. Satula and R. Wyss, *Nucl. Phys. A* **676**, 120 (2000).
[198] N. Schunck, ed. *Energy Density Functional Methods for Atomic Nuclei* (IOP Publishing, Bristol, 2019).
[199] N. Schwierz, I. Wiedenhover, and A. Volya, arXiv:0709.3525. *Parameterization of the Woods-Saxon potential for shell-model calculations.*
[200] R.A. Sen'kov and M. Horoi, *Phys. Rev. C* **82**, 024304 (2010).
[201] R.A. Sen'kov and V.G. Zelevinsky, *Phys. At. Nuc.* **74**, 1296 (2011).
[202] R.A. Sen'kov, M. Horoi, and V.G. Zelevinsky, *Comp. Phys. Comm.* **184**, 215 (2013).
[203] R. Sen'kov and V. Zelevinsky, *Phys. Rev. C* **93**, 064304 (2016).
[204] E. Shuryak, *Quantum Many-Body Physics in a Nutshell* (Princeton University Press, 2018).
[205] C.H. Skiadas and C. Skiadas, *Handbook of Applications of Chaos Theory*, CRC Press, 2016.
[206] N. Skribanowitz, I.P. Herman, J.C. MacGillivray, and M.S. Feld, *Phys. Rev. Lett.* **30**, 309 (1973).
[207] F.T. Smith, *Phys. Rev.* **118**, 349 (1960); Erratum *Phys. Rev.* **119**, 2098 (1960).
[208] V.V. Sokolov and V.G. Zelevinsky, *Phys. Lett. B* **202**, 10 (1988).
[209] V.V. Sokolov and V.G. Zelevinsky, *Nucl. Phys. A* **504**, 562 (1989).
[210] V.V. Sokolov and V.G. Zelevinsky, *Fizika (Zagreb)* **22**, 303 (1990).

[211] V.V. Sokolov and V.G. Zelevinsky, *Ann. Phys. (N.Y.)* **216**, 323 (1992).
[212] V. Sokolov and P. von Brentano, *Nucl. Phys. A* **578**, 134 (1994).
[213] V.V. Sokolov and V. Zelevinsky, *Phys. Rev. C* **56**, 311 (1997).
[214] V.V. Sokolov, B.A. Brown, and V.G. Zelevinsky, *Phys. Rev. E* **58**, 56 (1998).
[215] S. Sorathia, F.M. Izrailev, G.L. Celardo, V.G. Zelevinsky, and G.P. Berman, *Europhys. Lett.* **88**, 27003 (2009).
[216] V. Spevak, N. Auerbach, and V.V. Flambaum, *Phys. Rev. C* **56**, 1357 (1997).
[217] M. Srednicki, *Phys. Rev. E* **50**, 888 (1994).
[218] G. Sterman, *An Introduction to Quantum Field Theory* (Cambridge University Press, 1993).
[219] J.R. Stone and P.G. Reinhard, *Prog. Part. Nucl. Phys.* **58**, 587 (2007).
[220] D. Styer, *The Physics Teacher* **57**, 454 (2019).
[221] A.N. Sultanov, D.S. Karpov, Y.S. Greenberg, S.N. Shevchenko, and A.A. Shtygashev, *Low Temp. Phys.* **43**, 799 (2017).
[222] O.P. Sushkov and V.V. Flambaum, *Sov. Phys. Usp.* **25**, 1 (1982).
[223] A. Szafer and B.L. Altshuler, *Phys. Rev. Lett.* **70**, 587 (1993).
[224] A. Tayebi, T.N. Hoatson, J. Wang, and V. Zelevinsky, *Phys. Rev. B* **94**, 235150 (2016).
[225] M. Thoennessen, T. Baumann, J. Enders, N.H. Frank, P. Heckman, J.P. Seitz, A. Stolz, and E. Tryggestad, *Nucl. Phys. A* **722**, C61 (2003).
[226] M. Thoennessen, *The Discovery of Isotopes*, (Springer Switzerland, 2016).
[227] R.G. Thomas, *Phys. Rev.* **88**, 1109 (1952).
[228] E.J. Torres-Herrera, D. Kollmar, and L.F. Santos, *Physica Scripta T* **165**, 014018 (2015).
[229] R. Tsu and L. Esaki, *Appl. Phys. Lett.* **22**, 562 (1973).
[230] N. Ullah, C.E. Porter, *Phys. Lett.* **6**, 301 (1963).
[231] M.G. Urin, *Phys. At. Nucl.* **8**, 817 (1977).
[232] V. Vedral, *Introduction to Quantum Information Science* (Oxford University Press, 2006).
[233] A. Venugopalan, *Phys. Rev. A* **61**, 012102 (1999).
[234] J.J.M. Verbaarschot, H.A. Weidenmüller, and M.R. Zirnbauer, *Phys. Rep.* **129**, 367 (1985).
[235] M. Viana, *Lectures on Lyapunov Exponents*, Cambridge Studies in Advanced Mathematics, Vol. 145, (Cambridge University Press, Cambridge, 2014).
[236] A.V. Voinov, B.M. Oginni, S.M. Grimes, C.R. Brune, M. Guttormsen, A.C. Larsen, T.N. Massey, A. Schiller, and S. Siem, *Phys. Rev. C* **79**, 031301 (2009).
[237] A. Volya, B.A. Brown, and V. Zelevinsky, *Phys. Lett. B* **509**, 37 (2001).
[238] A. Volya, V. Zelevinsky, and B.A. Brown, *Phys. Rev. C* **65**, 054312 (2002).
[239] A. Volya and V. Zelevinsky, *Phys. Lett. B* **574**, 27 (2003).
[240] A. Volya and V. Zelevinsky, *Phys. Rev. C* **67**, 054322 (2003).
[241] A. Volya and V. Zelevinsky, *AIP Conf. Proc.* **777**, 229 (2005).
[242] A. Volya and V. Zelevinsky, *Phys. Rev. C* **74**, 064314 (2006).

[243] A. Volya and V. Zelevinsky, *Nucl. Phys. A* **788**, (2007) 251c.
[244] A. Volya and V. Zelevinsky, in *Fifty Years of Nuclear BCS*, R.A. Broglia and V. Zelevinsky (eds.) (World Scientific, Singapore, 2013), p. 73.
[245] A. Volya and V. Zelevinsky, *Yad. Fiz.* **77**, 1024 (2014) [*Phys. At. Nucl.* **77**, 969 (2014)].
[246] A. Volya and V. Zelevinsky, *J. Phys. Complexity* **1**, 025007 (2020).
[247] E.K. Warburton and B.A. Brown, *Phys. Rev. C* **46**, 923 (1992).
[248] H.A. Weidenmüller and G.E. Mitchell, *Rev. Mod. Phys.* **81**, 539 (2009).
[249] V. Weisskopf, *Phys. Rev.* **52**, 295 (1937).
[250] E.P. Wigner, *Ann. Math.* **53**, 36 (1951); **62**, 248 (1955).
[251] D.H. Wilkinson, *Phil. Mag.* **1**, 379 (1956).
[252] M. Wilkinson, *J. Phys. A* **22**, 2795 (1989).
[253] D. Wintgen and H. Marxer, *Phys. Rev. Lett.* **60**, 971 (1988).
[254] R.B. Wiringa, V.G.J. Stoks, and R. Schiavilla, *Phys. Rev. C* **51**, 38 (1995).
[255] S.S.M. Wong, *Nuclear Statistical Spectroscopy* (Oxford University Press, 1986).
[256] R.D. Woods and D.S. Saxon, *Phys. Rev.* **95**, 577 (1954). *Diffuse surface optical model for nucleon-nucleus scattering.*
[257] C.S. Wu, E. Ambler, R.W. Hayward, D.D. Hoppes, and R.F. Hudson, *Phys. Rev.* **105**, 1413 (1957).
[258] J. Zakrzewski and D. Delande, *Phys. Rev. E* **47**, 1650 (1993).
[259] G.M. Zaslavsky, *Hamiltonian Chaos and Fractional Dynamics* (Oxford University Press, 2005).
[260] V. Zelevinsky, B.A. Brown, N. Frazier, and M. Horoi, *Phys. Rep.* **276**, 85 (1996).
[261] V. Zelevinsky and A. Volya, *Phys. Rep.* **391**, 311 (2004).
[262] V. Zelevinsky, *J. Phys. G* **37**, 064024 (2010).
[263] V. Zelevinsky, *Quantum Physics*, Vol. 2 (Wiley-VCH, Weinheim, 2011).
[264] V. Zelevinsky and A. Volya, *Phys. Scripta* **91**, 033996 (2016).
[265] V. Zelevinsky and A. Volya, *Physics of Atomic Nuclei* (Wiley-VCH, Weinheim, 2017).
[266] V. Zelevinsky, S. Karampagia, and A. Berlaga, *Phys. Lett. B* **783**, 428 (2018).
[267] V. Zelevinsky and M. Horoi, *Prog. Part. Nucl. Phys.* **105**, 180 (2019).
[268] A. Ziletti, F. Borgonovi, G.L. Celardo, F.M. Izrailev, L. Kaplan, and V.G. Zelevinsky, *Phys. Rev. B* **85**, 052201 (2012).

Index